Technology and Social Change

Technology and Social Change

EDITED FOR THE COLUMBIA UNIVERSITY SEMINAR ON

TECHNOLOGY AND SOCIAL CHANGE

BY *Eli Ginzberg*

Columbia University Press   New York and London   1964

COPYRIGHT © 1964 COLUMBIA UNIVERSITY PRESS
LIBRARY OF CONGRESS CATALOG CARD NUMBER: 64-17158
MANUFACTURED IN THE UNITED STATES OF AMERICA

Preface

Since Professor Warner has set out in his introduction the background of the Columbia University Seminar on Technology and Social Change, which fathered this book, I can limit my remarks to matters of a more strictly editorial nature.

The task of preparing this volume for press was undertaken at the request of my colleagues on the steering committee of the Seminar, with the understanding that the major work would be carried out by my wife, Ruth Szold Ginzberg. She was the *de facto* editor. To the extent that the free-flowing discussions have form and content, the merit is hers. I retained, however, overall responsibility for the assignment. Our work was greatly facilitated by the excellent summary notes that Mr. Dean Morse, the secretary of the Seminar, had prepared from the tapes at the end of each of the five meetings.

The steering committee is deeply indebted to the five speakers, who took the time and effort not only to prepare illuminating presentations, but also to review and revise them. The editors, *de facto* and *de jure*, restrained themselves from stylistic alterations in their manuscripts.

Radical surgery was applied, however, to the discussions that followed each of the presentations. These were replete with strong opinions that were juxtaposed to equally strong but wholly contradictory opinions. It would have been ineffably boring to the reader to interlard the text with such statements as: "one discussant held," or "another discussant adduced the opposite view," or "a third believed that there was some merit in each of the positions previously advanced." The editors, therefore, decided to include almost every position that was advanced, without noting whether it represented an extreme point of view, had minority support, or could be said to reflect a consensus. The concluding chapter does, however, provide a clue to the center of gravity of the discussions.

A seminar, as Professor Warner makes clear, is really an extended dialogue. Therefore, much that is said originally is repeated later, sometimes in the same words, sometimes in different words. While some of the repetitive formulations have been thinned out, some have been left in deliberately to help the reader recognize the issues which proved of deep and continuing concern to the members.

The fact that a seminar is a dialogue also helps to explain what might otherwise appear to many readers as an anomaly. Quite contradictory points of view were presented, sometimes sequentially and sometimes intermittently. To have eliminated the conflicts and contradictions would have been to rob this report of the spirit of the discussions.

A few additional comments. The much greater than average length of Dr. DeCarlo's paper is explained by the fact that it was prepared specifically at the request of the steering committee to provide a broad framework for the succeeding discussions. The steering committee notes with appreciation the arrangements made by the U.S. Department of Labor—Office of Automation, Manpower and Training—to disseminate broadly copies of this report, thereby assuring that a larger group of interested citizens might be drawn into the vortex of these discussions.

ELI GINZBERG
Director, Conservation of
Human Resources

Columbia University
January, 1964

Contents

	Introduction, by Aaron W. Warner	1
1.	Perspectives on Technology, by Charles R. DeCarlo	8
2.	The Post-Industrial Society, by Daniel Bell	44
3.	The Aerospace Industry, by Earl D. Johnson	60
4.	The Dynamism of Science and Technology, by William O. Baker	82
5.	Productivity and Economic Growth, by Solomon Fabricant	108
6.	Confrontations and Directions, by Eli Ginzberg	136

Introduction

by Aaron W. Warner

Professor of Economics, Columbia University

The intellectual ferment and curiosity inspired by recent technological advances in industry have led to the opening of vast new areas for scholarly inquiry and research. Although scholars have long been concerned with technological change and its consequences, the more spectacular developments in the fields of electronics and automated processes excite the imagination and invite speculation. As one writer has aptly observed: "A source of great authority over nature, the modern scientific-technology promises to be both the hope of man's future and the instrument of his enslavement or his destruction. If we are to avoid the disasters it lays open to us and take advantage of the opportunities it presents, we . . . must understand what modern technology is, what it means, and what must be done with it if it is to serve man well."[1]

In a somewhat more pragmatic context, the increasing preoccupation of industrial nations with economic growth has centered attention on technology and innovation as the key to a more rapid increase in industrial productivity. The experience of these nations with the problems of growth over the past two decades has in turn led to a sharpened awareness of the widespread ramifications of technological change in the whole of society, and to a greater realization that the pace of change is

[1] Carl F. Stover, in *Technology and Culture* (Fall, 1962), p. 383.

dependent not only on innovation but on the adaptability of the economic and social structure into which it is introduced.

The fundamental nature of these problems, their appeal to many disciplines, and the controversies they engender, all suggest the importance of a suitable forum for the sifting of issues and for an exchange of views. Such a forum would also be useful in bringing together the various insights gained from experience and research in many disparate fields and placing them in new contexts for further refinement and exploration. It is primarily to these considerations that the Columbia University Seminar on Technology and Social Change owes its origins. The proceedings of the first five meetings of the Seminar, which form the basic chapters of this volume, are an attempt to frame the relevant issues and to lay the basis for future discussions and study.

For those who are unfamiliar with the Columbia University seminars, it may be appropriate at this point briefly to describe their nature and purpose. The seminars, of which there are now a substantial number, are set up as permanent, independent organizations within the University. Established on inter-disciplinary lines, they draw on the faculties of all departments of the University and provide the opportunity for the cooperative participation of scholars in many fields. The individual seminars are devoted to the study of the particular basic institutions or sets of problems in which their members have an interest. Thus, there are seminars which deal with such diverse problems as the State, the problem of Peace, the Renaissance, studies in religion and culture, higher education, the theory and practice of organization and management, and so on. In most cases, membership in the seminars also includes participants drawn from the faculties of other colleges and universities, government, business, labor, the professions, various foundations and research groups, and other organizations with related interests. As a result, the seminars provide the opportunity for a con-

INTRODUCTION   3

tinuing exploration of problems by groups of experts from different disciplines, many with a practical as well as theoretical grasp of their respective subjects.

The Seminar on Technology and Social Change, which was set up in the fall of 1962, is the most recent of the University seminars and follows this general pattern. Indicative of the focus of the new Seminar was the decision to explore the implications of technological change rather than the narrower subject of automation, which was regarded as merely one manifestation of a much larger problem. The responses to a questionnaire that was sent to prospective members prior to the formation of the Seminar showed clearly that it was necessary to have a very broad frame of reference. Indeed, as one respondent had indicated, the Seminar appeared destined to perform two major roles: 1) to document the effects of technology and to demonstrate its universal sweep, and 2) to discuss the social system with which the new technology would operate in coming years. At the same time, specific questions were proposed for investigation. These covered a wide scope, but the following major categories seemed to predominate:

a) A historical approach, which would put recent technological innovations in historical perspective, and isolate the likely future trends;

b) An inquiry into the nature and causes of technological change, emphasizing the role of research and development, education, the growth of scientific knowledge, and a broad range of other social and economic factors;

c) The problems of social adaptation to technological change, including the impact upon our major institutions and cultural patterns, and upon our democratic way of life;

d) The effects of technological change and automation upon the nature of work and work relationships, employment and unemployment, skill levels, consumption patterns, and the operation of the free competitive market;

e) The responsibilities of government and of leaders of industry to mitigate social costs inherent in the process of change.

The discussions at the first meetings of the Seminar reflect these diverse interests. To a considerable extent, this was the result of deliberate planning. The speakers for the first five meetings were selected to present a broad spectrum of ideas from vastly different orientations and vantage points. It was recognized that too board an approach would invite confusion; yet it seemed desirable not to narrow down the subject unduly at the start in order to allow scope for imaginative and creative formulations of the problems that were of prime importance. It was assumed that the Seminar would ultimately select the key issues for a more thorough investigation. This task remains as the next phase of the Seminar's development.

An introduction to the work of the Seminar is provided by the paper of Dr. DeCarlo. He speaks with philosophic insight of the profound effect of technology upon the world of today, and particularly of the plight of the individual in an environment which becomes increasingly technological and impersonal. His focus is directed to human values. In this setting, other problems fall into place. Who must provide leadership in the application of the new techniques? How can future leaders acquire scientific insights and still acquire the "fabric of our values and purposes"? Can man, by the development of his administrative skills, succeed in preserving and extending his freedom? Do the complexity of modern society and the omnipresence of large-scale organizations provide adequate opportunity for self-fulfillment? What in this respect are the responsibilities of large-scale organizations, including the government? Dr. DeCarlo also touches on a major source of the conflict in viewpoints concerning the future of our industrial society—the temperamental and vocational biases of those who make the predictions.

Professor Bell's paper analyzes the forces that are reshaping industrial society. In the society of the future, he states, intellectual pursuits rather than business activities will set the predominant tone, and the universities, the research institutes and corporations, and other intellectual centers will be the major institutions. The principal factors that give "a new character to society and . . . its problems" are identified as the exponential growth of science, the growth of intellectual technology, and the growth of research and development activities. The contribution of each is developed in a most stimulating essay. In response to the discussion, Professor Bell contends that in studying societies and cultures, insights are sometimes to be gained by singling out the new and extreme.

In the same pioneering spirit, Dr. Baker has explored the frontiers of science. His emphasis is upon the acceleration in the rate of scientific discovery and the significant narrowing of the interval between discovery and application. A concomitant development is the increasing impact of science as the processes of application become more sophisticated and complex. Dr. Baker also has much of interest to say about the generation of invention and the cost of innovation in a modern context. Particularly challenging is his conclusion that we must have systems research and systems development and systems engineering to utilize the new scientific discoveries, along with the necessary adaptations in our institutions to accommodate expensive, complicated, and concentrated systems techniques. As the discussion indicates, the inevitability of widespread systems requirements for technological advance is by no means accepted uncritically by other members of the Seminar.

Problems connected with the use of technical manpower in defense industries are taken up by Mr. Johnson. Increased emphasis on the most advanced technologies and the custom-built nature of the items covered by defense contracts have led to a steady shift in the ratio of scientific and engineering specialists to other types of production workers. This leads to a number of problem areas. What, for example, is the effect on the role of management when managerial decisions must depend to a large extent on scientific judgments? With the accretion of high-priced scientific talent in industry, what implications flow from the fact that the most productive years for high-level scientists appear to occur early in their careers? Under conditions in which there are rapid shifts in manpower requirements under defense contracts, what is the impact upon the morale of brilliant and highly trained scientists of job instability and long periods of unemployment? Mr. Johnson also raises questions concerning the significance of accelerated technological advances in the aerospace industries for the civilian sector of the economy. Will the technologies, processes, and products generated primarily for defense purposes be transferred to other civilian uses? To what extent is this likely to occur? What problems will it raise for our economic, governmental, educational, and social institutions? Mr. Johnson's paper and the discussion suggest the danger of extrapolating too much from the experience of the defense sector of the economy. It remains to be demonstrated that the experience of other sectors will parallel that of the defense sector. Yet the drastic effects of the technological changes in this area cannot be overlooked.

Dr. Fabricant's presentation is devoted in large part to clearing up a number of misconceptions concerning the sources and rate of growth of productivity in our economy. He considers the proper measurement of productivity, the cyclical character of productivity gains, and the factors other than technology which affect the growth of the economy. He also explores the intricate relationship between productivity and employment or unemployment. With his emphasis on the complexities of measuring productivity change, its cyclical character, and its long run tendencies, he in effect challenges the more intuitive apostles of the theory of revolutionary change to prove their case.

A summation of the major themes running through the discussion, with an eye to the development of future Seminar programs, is provided in a concluding analytical section by Professor Ginzsberg. In sifting through and classifying the complex assortment of concepts and problems, Professor Ginzberg indentifies the major areas of agreement and disagreement among the Seminar participants and assesses the extent to which consensus and clarification were achieved. The major areas in which questions were raised are identified as the key factors underlying technological change, the rate of technological change, productivity and economic growth, and social change. Each presents its own range of interrelated problems. Among the more challenging issues suggested for future discussion are the complex relationships between science and technology, the dangers implicit in the heavy concentration of scientific and engineering talent in the defense-space sector of the economy, the implications of the vast expenditures by government on research and development for private initiative, the adequacy of existing measures of productivity and economic change for determining the impact of modern technological developments, and the serious implications of technological change for education and training. The essay succeeds in skilfully fitting the Seminar discussions into a uniform pattern leading from the scientific underpinnings of technological change to the measurement and evaluation of its economic impact, and finally to the adjustments in our cultural and social institutions which technological change will inevitably produce.

The Seminar has thus opened a Pandora's box of issues. It has been guided by a small steering committee, which we hope will act as a research group to assist it in resolving some of the questions, checking on the findings of other scholars interested in similar problems, and developing ideas for additional research. Basically, however, the forte of the Seminar will undoubtedly remain discussion, with full scope for creative thought.

Perspectives on Technology

1   by Charles R. DeCarlo

Director of Education, IBM Corporation

While there is a wide divergence of opinion as to the magnitude and rate of technological and social change now under way, it is generally agreed that the world in which the American finds himself today is altered substantially from that in which his grandfather existed. This has been referred to as "the world of absolute possibilities." Whether or not this is true, we can discern the profound effect of technology upon our lives.

First, we have seen science and technology applied toward increasing the volume of goods and services we use. In fact, some of our major problems today arise out of the new possibilities inherent in the coupling of the production, distribution, and consumption functions within our economic society. Second, there has emerged, as a result of the increasing influence upon society of technological determinism, what appears to be a genuine concern for the role of the individual and individual freedoms. Finally, there is the fact of America's deep international involvement in a world made small by the collapse of space and time, and fraught with ever-present concern over the possibility of absolute destruction.

These environmental factors, all depending primarily upon technological development for their existence, constitute a set of background patterns which are continuously altering and shaping our institutional fabric and determining our future.

One can be optimistic or pessimistic as he faces this future, hopeful or despairing as he considers the question of individual life; he can feel menaced or confident as he ponders the use of science and technology, depending upon his background, philosophical beliefs, present responsibilities, body chemistry, and emotional make-up. This chapter presents the views of a businessman, optimistic that the institutions of pluralism can be made to work in an open society, and firmly believing that there is ample room in technology and bigness for the individual to experience existence, with all this implies. This is not to suggest, however, that ours is the best of all possible worlds and that the future will take care of itself. Nor is it to suggest that some other time in the past was better than this and it would be well if we could return to a more simple "humane" world.

Jacques Ellul writes: "One may well regret that some value or other of the past, some social or moral form, has disappeared; but, when one attacks the problem of the technical society, one can scarcely make the serious claim to be able to revive the past, a procedure which, in any case, scarcely seems to have been, globally speaking, much of an improvement over the human situation of today. All we know with certainty is that it was different, that the human being confronted other dangers, errors, difficulties, and temptations. Our duty is to occupy ourselves with the dangers, errors, difficulties, and temptations of modern man in the modern world. All regret for the past is vain; every desire to revert to a former social stage is unreal. There is no possibility of turning back, of annulling, or even of arresting technical progress. What is done is done. It is our duty to find our place in our present situation and in no other. Nostalgia has no survival value in the modern world and can only be considered a flight into dreamland." [1]

---
[1] Jacques Ellul, in *Technology and Culture* (Fall, 1962).

Our ability to "find our place in our present situation" depends upon and is conditioned by our views of the world which we as individuals inhabit—our view of the physical world, that is, the world of things, our view of the world of human relationships, and our view of the individual nature of man.

We are today dominated by our past philosophies and institutions, which above all recognize the importance of science and technology and which are in turn altered and reshaped by the lusty growth in every dimension of the wondrous world of "mechanism." The modern world, which experienced the first industrial revolution and which, according to some, is in the throes of a second, is a small segment of recorded history. Its roots go back to Descartes, Newton, and Bacon—each with their "rational" view of a world when order is supreme. During the last two and a half centuries our view of the physical has been predicated upon the Newtonian "world machine"—which assumed that the operations of the world follow immutable laws, capable of quantification, and enabling man, through his reason, continuously to expand his control over nature.

In this view, the world is literally computable, if the given state is known and the laws revealed. In the order of things, as science reveals and exploits the universal laws, life will be made better. In the words of Bacon, "the goal of the sciences is none other than this: that human life be endowed with new discoveries and powers." Coupling this thought with the doctrine of the perfectability of man, many men concluded that all changes taking place were progress. Thus, the eighteenth- and nineteenth-century science tended to support the view that world affairs follow an upward spiral, orderly and predictable, in which human perfectability and technical progress intertwine. Much must be said for this view, which Bacon articulated, since it led to the control of energy (at least on a Newtonian scale), the development of mechanical engines, and the application of mechanisms to a point where they can perform human-like functions. Thus we see a rising amount of energy available to peoples in the world, the organization and specialization of work, and the beginning of modern urban civilization.

Radiating from this development came the market economy, the political institutions, and the individual values of our current society.

But in the twentieth century we have the new physics which postulates indeterminacy as a prime condition and uses statistical viewpoints to predict future states. Application of these ideas has led to unlocking secrets of those enormous forces heretofore considered primal, undisturbed, and locked within the stuff of which the Newtonian world was made. Such a development must have profound effects, coming as it does at a time when the age of technical progress seems to have begun to achieve full development.

We must remember that the essence of scientific rationalism is a belief in objective scientific truth. The "real" properties of the world are those which can be quantified, measured, and made susceptible to mathematical formulation. The objective reality moves inexorably onward, pursued by scientists in search of its relationships and measurements and providing a mirror in which much of our idea of man is structured in the machine's likeness. For in scientific rationalism the subjective world of feelings, values, and the many qualitative aspects of life not susceptible to measurement or mathematical manipulation is considered a separate and "imperfect" aspect of the human mind.

Barrett has remarked on this dualism—and the significant split of the human and scientific world it forces. "The Cartesian era of mathematical physics now approaches its violent climax . . . with all the human turmoil of our period, with its political unrest and individual rootlessness, we are aware of the skeleton that lurks in the Cartesian closet: our power to deal with the world of matter has multiplied out of all proportion to our wisdom in coping with the problems of our human and spiritual world." [2]

Much of the philosophic and artistic commentary of today concerns this very problem. Existentialism, modern art, and the Theater of the Absurd probably could not have come into being had not the control of the physical environment made possible that kind of world in which man's encounter with man, rather than the drudgery of work and meniality, become background for the human condition.

This is not the heroic encounter found in classic literature, where the individual protagonist attempts an imposition of his will on the world in opposition to others and where the world, physical or imagined, largely defines character and conflict. The modern encounter is more of man in a desert—where articulation and control of technology has removed much of the definitive relationship between man and environment which gave drive to the former struggle. There is created an aloneness of the modern protagonist which forces an emphasis upon, and demand for, "communication" and mutual understanding of the different roles the individual plays in a world in which large human organizations also move as viable and dominant personalities. This is demonstrated in the plays of Beckett with their barren settings and language which make minimum reference to physical reality, or in the works of Genet with their poetic reflection and refraction of roles assumed by individuals in a fantastic and impersonal world.

In the tragic view of life it is the individual, proudly exalting his fatal flaw, who faces Fate and suffers a fall. In the modern

---

[2] William Barrett and H. D. Aiken, *Philosophy in the Twentieth Century* (New York, 1962).

world it is more likely the large human organizations which shape the destinies of our time and through or against which the individual must establish his roles.

The individual's concern for personal and biological autonomy within the collectivist personality is one of the principal features of our modern tension. It appears that the larger personality exists to guarantee freedom *from* such things as hunger, cold, fear, and warmth. However, the individual autonomy has an inchoate demand for freedom *to* think, speak, feel, to exercise choice, and to know who you are when alone. This tension between the larger and the individual personality could not exist if it were not for the advances in our science and technology which have enabled the larger institutions to grow in providing these minimum levels of freedom *from*.

Within today's seemingly autonomous organizations—the "new machines" of science and technology—the very presence of technology imperceptibly alters our view of reality, constituting as it does a direct influence upon our senses, accumulating by its presence what appears as an ability to control our future, and making us increasingly independent of physical events and interdependent for personal relationships and values. Here we see an increasing evidence of the effect of the impersonal technological world in creating the need for a new "personalist" philosophy.

While at first blush it appears that such viewpoints and questions are pessimistic, they are vital in that they are antidotes to both the alarmist *and* enthusiast about technology, for they call attention to the necessity for reviving and revitalizing the concept of the individual existence in our world.

It is generally agreed that the cornerstone of our modern technological society is information. Peter Drucker has coined the phrase "knowledge workers" to describe the activities of people who develop, direct and administer our control of technology. This dependence upon knowledge brings a new importance to those two principal agencies concerned with knowledge—the universities and the scientific community. The increasing involvement of science in the creation of technological gadgetry is making it difficult for scientists to provide criticism based upon their specific competencies. The relatively limited amount of public discussion on space programs, atomic energy, and CBR warfare, are indicative of the difficult transition period through which we are going in learning how to mesh the role of professional specialization into the process of our political decision making.

Speaking of the broader influences of the university community, and the effect of scientific revolution, Justice Douglas writes, "Those who finance the scientific revolution usually control those who work for them. The impact of this control on our universities is so great that their autonomy is threatened.

"There is ample evidence that our colleges and universities have become citadels of anti-radicalism. One who sits on the sidelines and listens to the highly factual, unimaginative, and tranquilizing essays they produce sometimes concludes that he is a witness to an inquest. Yet if planning in this fast-moving age of technology is not being designed in university circles, from what source will it come?" [3]

Great pride is expressed in the historic mission of the university to preserve, develop, and transmit knowledge. In times when the power of knowledge was exercised over longer time spans and subjected to natural dispersion by distance and smaller populations, the universities could play their leading social role in preparing the raw material for future leadership. It seems today, however, that there is much opportunity to expand the vital role in our society of intellectual leadership

[3] William O. Douglas, "Freedom of the Mind," American Library Association.

and criticism as an antidote to our ability to communicate automatically massive doses of the trivial.

The modern treatment of information has greatly changed the world in which we live, through the transmission of data, pictures, and ideas at high speed over great distances to millions of people. Through mass communications there can be transmitted and communicated quickly complete ranges of ideas, feelings, and attitudes. This technological phenomenon has placed great power and responsibility in the hands of those institutions which were historically responsible for, and responsive to, the dissemination of ideas. Possibly because of the development of the market economy, and the trend for centralization in our population in response to the technological demands of production, there has grown a tendency for the number of communication channels to be increasingly limited to fewer and larger organizations. Recent developments in educational television, and the few hopeful signs that universities will take an active interest in what is communicated, as well as how effectively the technique works, are encouraging signs. Perhaps in the future we will have that healthy dissent to which Justice Douglas referred when he wrote: "The safety of the Republic lies in unlimited discourse. Only when the mind is free to explore problems to the horizon is man free to challenge and criticize intelligently those in power and summon an opposition to depose them." Speaking further of the importance of communication and information in the realm of ideas, he points out that while tyrants may suppress men, "Ideas are more dangerous than armies. Ideas have immortality, ideas cross impassable frontiers, ideas penetrate any Maginot line of conformity. Voices can be stilled, men and women imprisoned, books burned, but their ideas live on to torment the executioners, jailers, and censors."[4]

[4] Douglas, "Freedom of the Mind."

There are several possibilities for future technological development. First, it seems that we will be able to extend the present ability of machines to perform processing work. At present an automatic computing system does one thing at a time. Its work flows sequentially no matter how fast the individual elements function. By making the heart of the machine, its timing element, go faster and faster, i.e., many millions of times per second, it is possible for the system to accomplish much work in time scales commensurate with the capability of the human brain. We may think of it as accomplishing work by connecting long chains of decisions (or wires and electronic elements if you prefer a visualization). However, as it performs work it is always on one track in time. As the system proceeds to thread its decision chain in the maze of possibilities, it has available to itself at each point a limited amount of the data and instructions for determining the future of the decision path. Because each point in the decision chain represents the triggering of a few primitive elements, analogous to the operation of our own nervous system, we have to devise methods and languages capable of expressing our complex and purposive problem into expressions which such a decision path might encompass. Because of this great gap between our purposes and the machine's nature the work done to date upon machines has depended heavily upon mathematical or logical formulation, or the building of logical analogues to physical systems. How much of human purpose and desire lie above and beyond the world of machines!

Considering the nature of our computing systems today it appears highly probable that we can expect developments along the following lines:

1. Methods will be found to organize the functioning of machines so that they can achieve simultaneity through the specialization of work. Certainly in the case of our own body we

perform many functions at different levels of conscious control through specialization.

2. We will be able to provide much larger memories to enable the systems to bring more experience and data upon the operation of the decision-making path.

3. We will be able to build systems in which the decision path will not depend rigidly upon point by point connection of decisions (or wires and electronic components). Rather we can think of the system as being composed of a random collection of decision points and elements, with multiple cross connections and combinations. It is possible to conceive of such a system where certain input stimuli will cause specific output responses which can be appropriately reinforced. In this sense it can be said that the system learns.

It is unfortunate that we have not better language to describe such a system for there is immediately raised the question of equivalency between machine "learning" and "thinking" and the human functions. While this idea has given science fiction writers and certain popular columnists grist for their mills it seems trivial—an erroneous concern which springs from the comparison of the individual being to a system. When the machines demonstrate consciousness and multiple-purpose behavior, then perhaps we can see them as a competitive threat. This is not to assert that there is no danger inherent in the application of automatic systems to accomplish man-determined ends. The present status of air defense and nuclear deterrence is ample proof that the time response and magnitudes of energy involved in the automatic system must make us ponder, in this case, the reasonableness of "our man-determined end."

A second major area of future development lies in the ability of automatic systems to be connected and communicate with a far ranging environment. We know today that the eye, the ear, the nose, the tongue. and the hands are each in their own way limited receptors of the world around us. Arthur Clark has observed: "There are some senses that do not exist, that probably can never be provided by living structures, and which we need in a hurry. On this planet, to the best of our knowledge no creature has ever developed organs that can detect radio waves or radioactivity." He goes on to suggest that "In a crude way—and one that may accurately foreshadow the future—we have already extended our visual and tactile senses away from our bodies. The men who now work with radio isotopes, handling them with remotely controlled mechanical fingers and observing them by television, have achieved a partial separation between brain and sense organs. They are in one place, their minds effectively in another." Viewing the evolution of man and his earliest uses of tools, he conjectures a future evolution resulting from the synthesis of man and machine—in which the machine ultimately discards its purely organic component.

It is interesting that the burden of his exposition depends upon the awesomeness of space and our need to adapt in order to conquer it. In his words, "If we reduce the known universe to the size of the earth, then the portion in which we can live without space suits and pressure cabins is about the size of a single atom. . . . Like many other qualities, intelligence is developed by struggle and conflict; in the ages to come, the dullards may remain on placid earth, and real genius will flourish only in space—the realm of the machine, not of flesh and blood." [5]

It seems that he forgets that it is we who perceive the size of the universe and for whom, in the words of the poet, "All objects in the universe converge and we must find their meaning." Yet there is some reason to the position from which he extrapolates. It seems reasonable to expect that in the future machines will be made adaptable to their environment and

[5] Arthur Clark, in *Industrial Research* (November, 1961).

perform many of the tasks which represent human toil and drudgery today. But considering his long range and fantastic predictions, how poignant it makes our concern for the individual life of a child, or our distress at the ignorance or inequity which still persists throughout this dull and "placid earth."

Finally, in the future we will probably solve the problem of transmitting high energy over great distances efficiently. While there is no evidence which establishes the direction in which this might be done, it seems that it must come.

These possible future developments in science and technology underscore a major problem of our times—the question of who in society must provide the leadership in the application of new techniques. At the base is the problem of communication between technicians and scientists responsible for technological development and that leadership in our society responsible for the many institutions and organizations which constitute our ongoing expression of aims and purposes. The problem transcends the question of the "two cultures" propounded by C. P. Snow. Surely our present leadership can be made technically literate, and be given the necessary scientific insights so that they can exercise the choices required in the application of technology to our human activities. But going beyond this is the question of building future leadership and transmitting the fabric of our values and purposes to the oncoming leadership. With all the emphasis placed upon the importance of raising the number of engineering and scientific graduates in the country for present and near future needs we are apt to lose sight of this important fact. We know that the engineering schools are attempting to include certain of the "humanities" in the curriculum. In addition there is considerable feeling among scientists today that the traditional departments and the "humanities" in our educational system are not receiving appropriate support, either financially or through the availability of the top quality talent. Although we are probably over-emphasizing engineering and scientific training as our principal component of need, this is not to suggest that we diminish this activity but rather that we find appropriate agencies by which we can massively strengthen those other parts of the university community which provide the complete spectrum of human knowledge and wisdom.

But our concern goes beyond the formal training of the student, or the conflict of academic departments. It concerns that point in a man's career when he becomes a leader. That is when he must learn to do his job, but also has to relate it to a larger context in providing direction for others. Traditionally this ability has come out of a continuing experience, a set of relationships and attitudes which are demonstrated between himself and those who pass on the responsibility to him. It seems that one of the most subtle and profound effects of technological change can be that it constitutes a kind of discontinuity in this training for leadership. The question is how do we guarantee in a rapidly changing technical world the apprenticeship of wisdom that normally flows continuously in the transfer of leadership. This is particularly meaningful when technological and language barriers separate the man who leads and the man who must be trained. Too often the leadership feels estranged by the differences in experience and training. Questions of competency in technology and application may translate themselves into fear. In this situation leadership has difficulty in transmitting its values and experience and it often polarizes the more inexperienced in human affairs into using the barriers between them to reinforce their separation.

The success of our present leadership in responding to this challenge will effectively determine the nature and effect of future technological and social change. This is true in such diverse human activities as the use of systems in the teaching environment, the use of systems as planning instruments, the control of our military investment, and the solution to a host of political, industrial, and academic affairs.

When one surveys the books, papers, and testimony available on the subject of technological change, a wide divergence of views is apparent. Indeed there is a whole spectrum ranging from enthusiastic belief in to despairing pessimism about the future of our technological society. One is also impressed with the fact that the computer—or system, involving the computer as an element with feedback—becomes the principal focus of much of the debate. It appears at times that this aspect of our technological creation seems unduly emphasized when actually it is but one part of the total pattern woven by electronics, communications, and power engineering.

For the purpose of this discussion we will split the spectrum of viewpoints in three major segments. The first concerns those we will call the operational management of our society, the second are the intellectuals (economists, political scientists, and others), who act as critics of this system, and the third are those scientists and specialists who are participating in creating the technological change.

It seems that the leaders of business, labor, and government have the principal operating responsibility for our society and have today many more similarities in their viewpoints than differences. The present leaders naturally attempt to use the past to present the image of the future, conserving the values and attitudes which have made present institutions and organizations successful. They search for ways in which the present technological changes can be viewed as normal evolution. The problem, as seen through the eyes of our leaders, is to find agencies for our present imbalances, to provide stimulus to the already existent institutions, and to take care to provide the balance between change and stability—in such a way that our basic economic and political systems suffer minimum change. Their approach in general is to seek out adjustments, to intensify or weaken various operative forces, and to make present institutional formats work with minimum change. This is the conservative view and it guarantees that the future will preserve to the maximum extent the shape of those ideas which guaranteed success in the past. As A. N. Whitehead has pointed out: "The art of progress is to preserve order amid change and to preserve change amid order." In this statement is summarized the principal imperative for the leaders of our large organizations today.

Second, there are a host of concerned critics who provide a different viewpoint, both larger in scope and against a longer historical time scale. By and large it is their thesis that the world has changed sufficiently in the past several decades that we should reshape some of our institutional forms and relationships to meet the future confidently. Typical among this range of viewpoints are the following:

1. As a result of technological change we have expanded our ability to produce goods and services so radically that the whole basis of our present economic theory—scarcity—is replaced by a condition of potentially unlimited abundance. If true, the derivative effects of this upon our political system would be considerable.

2. Power, authority, and ownership have a changed relatiomship in the new technological environment. This is of particular concern as the public, private, and nonprofit forms of corporaite endeavor move to new dimensions in size, scope, and influence in the society.

3. The roles of planning, control, and leadership have a new importance in an environment increasingly technologicallly

deterministic. The proper use of the instruments of communication and computation in the legitimate functions of government constitute a genuine concern.

4. There exists the concern that as we grow the disadvantage that exists for those with lesser education will increase, giving rise to new class distinctions. The effect of education or its lack in this regard throughout the world is similarly a growing concern.

5. The shaping of world events creates new dimensions in the differences of rich and poor throughout the world. The possibility of our advanced technological society becoming rigidly affluent and inflexibly attached to the status quo is viewed as a principal danger in world affairs.

Generally speaking the viewpoints in this part of the spectrum are predicated upon optimism in that they hypothecate that our system, if sufficiently flexible and humane, can adapt itself to the future needs of the people as well as our role in world affairs.

Finally we have the viewpoint of the specialists and scientists. It is their general position that the advent of nuclear weapons, of high-speed computers, and rapid communication have created a totally new environment. They believe that our technological change to date constitutes a discontinuity and that most of our past values concerning work, leisure, ethics, and meaning will be profoundly affected by the new world. They perceive present and future technological environment as possibly menacing and capable of exceeding the control of its human leadership. Their general conclusion is that man must adapt himself to the new environment by controlled and rational choices. Theirs is a perfectionist viewpoint which requires the reshaping of the human organization to accommodate the flow of material reality. Implicit in their view of the world is the belief that an elite leadership should be developed to provide the controlling and planning elements for our future society. Such a viewpoint constitutes a type of inverted utopianism, for it suggests that if we could only change our natures we could live in a world wherein the coupling of technology and people, under proper control, would lead to a world of unlimited material well-being. Like most utopians they underestimate the importance of the values by which man presently lives, and are unable to provide the bridging mechanisms which would enable us to get from our present situation into the world of the future. Further, there is no indication that human nature would be capable of living in a world in which tensions, problems, and conflict were removed.

At the heart of their thinking is a profound pessimism about the nature of man. In light of the material and communication worlds in which so much order can be imposed, they find human variability and the desire on the part of each individual to have some disorder and detachment from the material world as frustratingly nonrational.

As our technological environment has become more widespread, we have come to depend upon the large organization in carrying out the tasks of our society. The management of these organizations in such a way as to provide successful accomplishment of mission, as well as individual freedom, is a major problem of our times.

Several years ago Harlan Cleveland wrote: "It is a measure of the national mood that at the peak of American power we should be seized with the worry that large-scale organization is somehow a Bad Thing—that the very administrative skill which enabled us to build this strength and brought us free-world leadership is itself a threat to freedom.

"My thesis here is the reverse: It is precisely by the development of his administrative skills that Man preserves and extends his freedom. The complexity of modern society and the omnipresence of large-scale organizations not only provide an opportunity for the fullest development of the responsible self; they actually place a premium on the exercise of a greater measure of personal responsibility by more people than ever before." [6]

The validity of this thesis depends upon the way in which the leadership of a large organization discharges its responsibilities with regard to the individuals constituting the organization, the intrinsic mission of the organization, and the relationship of the organization to others in the society. Indeed it is safe to say that these three areas of consideration lie at the heart of managing transition in this period of change.

Technological growth, geographic shrinkage, and new time scales have fostered the rapid development of new organizational problems. The extreme example of such growth is our Defense Department, which uses about 10 percent of our gross national product, spends more than the combined national product of Canada, Japan, India, or China, more than all the state and local governments in the United States, including all expenditures for public education for almost 50 million people from kindergarten to state universities. The existence of an organization of such size and far-reaching effect could probably not have been predicted as short a time as twenty years ago. In a similar way the rapid growth of business organizations, universities, the agencies of government, have exceeded what an observer in the late 1920s or early 1930s might have anticipated. We have seen such a range of new organizational forms develop that today we have the complete spectrum from the privately held business enterprise through nonprofit businesses, foundations, to new governmental commissions and agencies.

With respect to the mission and purposes of the organization,

[6] Harlan Cleveland, in *Saturday Review* (Feb. 28, 1959).

largeness tends to develop certain qualities of maturity. As an organization grows there is a tendency to develop basic policies within the organization which seek to minimize risk, to attempt stabilization of the environment, to attempt plans and controls for the future, and to develop predictable attitudes and routine practices. Just as the adult seeks security and control over his environment with measured consideration and the wise desire to use the past to guide the future, in contradistinction to the playfulness, curiosity, and risk-taking attitudes of the child, so the large organization must differ from the small. This means the organization tends to become locked into ongoing operations, plans, and attitudes, all built around utilizing ideas which have succeeded in the past. We have seen that with an excess of this attitude certain industries have missed the opportunity for applying new technology, or gathering new market opportunities, because they represented either too much change or risk measured against past practices.

The second aspect of largeness in organization, resulting from the technological opportunities available, concerns the role of the individuals constituting the larger personality. The first major consideration is the nature of the work performed by the individual. Largeness and technological change imply a specialization of work, removal of many repetitive menial tasks, the rationalization of planning and control, and the formalization of the social environment in which the individual participates in the organization. In some way each of these reflects the fact that the internal environment of the organization has a tendency to be atomistic, technically determined, and with growing emphasis upon efficiency.

Now it must be noted that because of the advance of technological developments the actual content of work is increasingly concerned with the handling of information and knowledge rather than physical process. It should be further noted that many of the attitudes, intellectual constructs and measures for the organization derive from the historical past when the ratio of muscle labor was in a much higher proportion to brain labor than today. Further, it is probably true that certain jobs, including some of those management tasks which Professor Melvin Anshen refers to as "programed decision making," will require less skill in the middle of the work spectrum than they do at present. This growth of the administrative, rather than the manual, nature of work must be recognized by the organization leaders. The difficulties of measuring the productivity and efficiency of clerical operations, engineers, scientists, salesmen, accountants, are well known. The old concept of scientific management in which the worker's efficiency was measured in terms of his material production cannot be applied to measure productivity and efficiency in today's increasingly "information-dependent" environment. However, instead of this being a burdensome imposition upon leadership, it should be an opportunity for creative human management. Because we are now concerned less with the building of machines and things we can turn our attention to the building of human organization to achieve the purposes and fulfill the needs of organizations within the community. This is the true meaning of the concept of the "profession of management." This profession is not now, nor will it be in the future, an easy activity. The tendency to use old and easier methods of measurement, to express the human problems in terms of the production and material values of the organization, the human desire to retreat from broader responsibilities, are all tensions working upon the modern manager in a large organization.

One aspect of this concern for "efficiency" in administration of organization has caused Kenneth Boulding to write that "playfulness, informal communications, and even extravagance and wastefulness in 'normal' times give an organization survival value in times of crisis, for they develop and keep open spare channels of communication and reserves of energy which can be drawn upon for 'serious' purposes in time of crisis. The greatest threat to survival can easily be efficiency." He seems to be telling us that the human energy, the latent sense of purpose and belief in the organization's aims and ideals must be wisely nurtured in good times so that they may respond in times of challenge.

Because we are an idealistic and pragmatic society we have responded in a very humane way to many of the situations outlined above. Certainly in the process of organization there is a tendency on the part of the organizational leadership to provide, and its constituent members to expect, increased security. Previously we have referred to the larger organizations' ability to guarantee the freedom *from*. Certainly today a look at the large American organizations, corporate, labor, university and government, amply demonstrates that it is the large organization which is guaranteeing, or attempting to guarantee, a measure of this security to an increasing number of people. Of course, one aspect of this phenomenon is the problem of who shall provide similar security and "freedom *from*" for those individuals who are not a part of large organizations. There is also the problem of the role and responsibility of organization in those times of change and dislocation when individuals must be able to move from one job or one organization to another. The President's Labor-Management Report speaks very well to this issue:

"The mobility of workers is reduced by factors running contrary to the demands of a dynamic society, and an economy in transition.

"The non-transferability of pension, seniority, and other accumulated rights may result in an employee's being dependent upon his attachment to a particular job as the sole means of protecting his equities.

"Desirable and essential mobility is affected by reluctance to leave home—because of personal ties, or because other members of the family may be working; by the cost of moving and possible losses on local property; and by the insecurity of jobs in a new locality."

A final word on the individual and the organization concerns his relationship to his work. Increasingly we see the individual becoming less craft oriented and more capable of adapting to new work environments. In a sense this means that he views his work as not being determined by "job," but rather by his loyalty, affiliation, and dependence upon the organization. This dependence can have the effect, in the long run, of bringing about the worst of those aspects referred to in the "organization" society.

However, the individual does have one important degree of leverage and freedom which enables him to contend with the organization—his training, education, and intellectual competence. For if one thing is evident, it is that the future will belong to those who study for it and continue an attitude of lifelong learning, for this will guarantee them the ability to shift from task to task within the organization or among organizations.

In summary, it appears that the effect of technological change has been to encourage the growth of largeness in organizations, the specialization and change in the nature of work, and the shift of our attitudes on the importance of education. The future progress of our organizations and our society in a large measure depends on the attitudes of our leadership. If there is an insistence upon the view that change affects only the material and technological, and that all past practices can fit easily into the new and larger world, then indeed we are headed for trouble. However, the excitement and opportunity inherent in the new mobility generated by education and intellectual competence, plus the latent sense of service at the heart of American idealism and pragmatism, will be mustered to take a new grasp on our problems and convert them into a better and more humane world. Whitehead has remarked that "The diffusion of literacy and average comfort and well being among the masses, in my opinion, is one of the major achievements in human history. With all its limitations, life in America is better and kinder than anywhere on earth I have ever heard of." There is too much in our fabric of values, our willingness to change, for us to accept a despairing or defaulting attitude on the part of our organizational leadership.

Perhaps the area where the phrase "technology and social change" has the most compelling urgency is that which concerns the government's use of technology. One observer has remarked, "Our democracy has ingested science, but it has not yet digested it—a measure of the infancy of our scientific society." We are made aware, almost daily, of the size and the importance of our defense budget. The fact that we have been "compelled to create a permanent armaments industry of vast proportion" moved President Eisenhower to sound the now famous warning in his farewell address.

There is reason to question the thesis, however, made in Fred Cooke's "The Warfare State," that a determined and willful clique of the military and industry are indirectly controlling our economy and keeping us in a state of cold war. Such a notion is totally out of character with our nature and the type of institutions we have inherited. The existence of such men in any considerable number would imply a discontinuity in American values far greater than those discontinuities which perplex us in the scientific and technological sense. Yet the fact remains that the role of science and technology and the machinery of government are changing slowly to accommodate the new facts of American life.

Two aspects of change related to technology in government merit brief comment. These are its effects on government decision-making apparatus, and the role of research and innovation in our economy.

Under the twin pressures of the cold war and the relentless pace of science and technology, scientists have increasingly entered into our national decision-making process. They appear as advisers to major branches of the executive departments, as members of commissions and councils, and as operating executives in our research and development apparatus. That such participation is motivated by the highest ideals of patriotism is surely above suspicion. But as Don K. Price points out in his article, "The Scientific Establishment," many scientists come peculiarly ill-equipped for their roles in national leadership. This is due to a general pride in scientific morality, a distaste for the compromises and accommodations of politics, and an ingrained desire to work in the open with judgments and results available for criticism by a community of peers. Mr. Price describes well what must be the resultant personal dilemma:

"To one who believes that science has helped to liberate man from ancient tyrannies—who, in short, still takes his political faith from Franklin and Jefferson and the Age of the Enlightenment—it is disconcerting to be told that he is a member of a new priesthood allied with military power. Yet the plain fact is that science has become the major Establishment in the American political system: the only set of institutions for which tax funds are appropriated almost on faith, and under concordats which protect the autonomy, if not the cloistered calm, of the laboratory."

The scientist who becomes enmeshed in the councils and administration of Big Science finds it a fast moving, complex, and tough environment in which the rules are more like the market place than the classroom or the laboratory. Describing the reality that exists, Dr. Hans Bethe writes:

"It is sometimes doubtful whether it is really a democratic process that is going on, especially because the agencies that have the money for scientific research—and now I mean military technological research—are precisely the agencies that want the maximum of military technological development. This includes both the armed forces and Congressional Committees. So, scientists who advocate development of weapons without restraint find a very ready public, while those who warn against the dangers of an unlimited arms race find a very hostile reception from many members of the Washington community."

Historically, the systems of checks and balances, political accommodations, and the multitudinous interests represented in the government have allowed us under the Constitution to adjust to change.

In the words of Justice Douglas: "The major achievement of the Free Society is in the ability to change the status quo without violence, to cast a current practice into limbo and adopt a new one by an election, to remake the economy or renovate an institution, yet not destroy it, to refashion even the structure of government by votes rather than by force." [7]

We do appear to be in such a process of "remaking and renovating our institutions." In spite of the complexity of technology, the size and pace of its investments, it appears that the Congress is awakening to and will grasp the new responsibility for providing its component of direction in national affairs affected by technology. In commenting on the $3.7 billion budget for NASA, Representative Brown is quoted as saying:

"There seemingly are few Members of the House . . . and I suspect very few citizens of this country . . . who know for a

---

[7] Douglas, "Freedom of the Mind."

certainty whether the amount contained in this bill is the proper one—we must accept this legislation on faith. We must accept them on faith and hope that the expenditure of these huge funds authorized in this bill will be in the best interests of the American people and the world peace we all seek." [8]

The fact that Representative Brown was candid enough to admit this is an encouraging omen. Half the battle will be won when this problem is clearly identified and articulated by a sufficient number of the politicians in the Congress. While the Congress may change slowly, it will change to demand more voice in the course of our future, for the legislative function is tough and resilient, springing as it does from the total variety, strength, and humanity of our people.

The second aspect of technology in government concerns its effect on our economy. Congressman Morris K. Udall points out that we spend approximately 75 percent of our budget for defense, diplomacy, and past wars; 20 percent for the general functions of government; and 5 percent for the functions of labor and welfare programs. It is his contention that the government is not growing out of proportion to our population and economy. With respect to health, education and welfare programs, he writes the following: "In 1939 we spent not 7 percent, *but 44 percent*, of our budget for labor and welfare programs. . . . In 1939 we spent *$30 per capita* on these programs. . . . In fiscal 1963, using the 1939 dollar to provide a fixed basis of comparison, we will spend *$16 per capita* for these same programs." [9]

His major concern is with the size of the defense and space budgets which divert much of the energy and national resource which might otherwise be made available for such things as massive education programs.

In another vein Mr. Gerard Piel has written on the role our

[8] Brown, in *Science* (Jan. 4, 1963).
[9] Morris K. Udall, in *New Republic* (Oct. 1, 1962).

defense budget plays in our national economy. Typical of his views is his contention that: "Military expenditure has taken up more than half of the federal budget and fully a quarter of our manufacturing output throughout this period. In the fiscal management of our economy, in other words, armament has played the same role as public works in the first two administrations of Franklin D. Roosevelt. After 10 years of this kind of pump-priming, is it any wonder that our magnificent industrial establishment should have burdened us with such an enormous surplus of weapons?"

But he goes on to assert that: "The arms budget is losing its potency as an economic anodyne. It is concealing less and less successfully the underlying transformation of our economic system. Progress in the technology of war, as in all other branches of technology, is inexorably cutting back the payroll. With the miniaturization of violence in the step from A-bombs to H-bombs, from manned aircraft to missiles, expenditure on armaments has begun to yield a diminishing economic stimulus." [10]

In spite of the pressures of cold war, the demands for military secrecy, and the involvement of so many interests in the process of defense procurement, issues this large and important should have the benefit of much public discussion.

There is another effect of the increase in federal expenditures for research and development. This concerns the responsibility for innovation and risk-taking in the development of new goods and services made available to both our civilian and government economy. Speaking of the new industries and new situations which have developed in this regard, the report to the President on government contracting for research and development says:

[10] Gerard Piel, in Phi Beta Kappa oration given at Harvard University, June 11, 1962.

"There are significant differences between these newer industries and others. While the older industries were organized along mass-production principles, and used large numbers of production workers, the newer ones show roughly a one-to-one ratio between production workers and scientist-engineers. Moreover, the proportion of production workers is steadily declining. Between 1954 and 1959, production workers in the aircraft industry declined 17 percent while engineers and scientists increased 96 percent. Also, while the average ratio of research and development expenditures to sales in all industry is about 3 percent, the advanced weapons industry averages about 20 percent and the aerospace industry averages about 31 percent.

"But the most striking difference is the reliance of the newer industries almost entirely on Government sales for their business. In 1958, a reasonably representative year, in an older industry, the automotive industry, military sales ranged from 5 percent for General Motors to 15 percent for Chrysler. In the same year in the aircraft industry, military sales ranged from a low of 67 percent for Beech Aircraft to a high of 99.2 percent for the Martin Company.

"The present situation, therefore, is one in which a large group of economically significant and technologically advanced industries depend for their existence and growth not on the open competitive market of traditional economic theory, but on sales only to the United States government. And, moreover, companies in these industries have the strongest incentives to seek contracts for research and development work which will give them both the know-how and the preferred position to seek later follow-on production contracts."

Considering that the federal government finances about 65 percent of the total national expenditure for research and development, and that the federal share is rising, it puts the following remark from *Business Week* in an even more important context.

"If research were considered an industry, it would rank among the top dozen manufacturing industries in the U.S. today. Already its employment (at 350,000) is close to three-fifths that of the automobile industry. By 1970, if current plans are fulfilled, the business of research, in dollars and persons employed, will push even higher into the top echelons of industry."

The net effect of these situations is that increasingly the federal government through the exercise of its defense budget will play a dominant role in the functions of innovation and risk-taking involved in technological developments. This may lead to a further blurring of the line between public and private organizations. The existence today of nonprofit corporations, foundations, and large university-managed research centers gives ample testimony that new institutional formats can be developed to accomplish these purposes.

The priority decisions which assigned the purposes of our national research and development programs should reflect the voice and responsibility of a larger segment of American leadership than merely that of the defense specialists. Several authors have pointed out our need to step up research and development in our basic industries as well as in civilian products and services. There is a great need here for creative application of product innovation, technological change, automation, etc., despite the present size of our national research and development effort. Speaking to this point, the Assistant Secretary of Commerce, Dr. Holloman says:

"Although the United States is spending about $16 billion yearly for research and development, 70 percent of this outlay is for government-sponsored activities. Less than $4 billion is being spent by industry for new products and processes, with only a third of that to improve productivity. Moreover, these industry research-and-development expenditures are heavily concentrated in a few companies, a few industries, and certain geographic areas.

"Other nations are devoting larger proportions of their national research-and-development expenditures and their technical manpower to stimulating the civilian economy. They are also enjoying higher growth rates in both income and productivity than we are. These nations, especially those in the rapidly growing European Common Market and Japan, are competing with us not only in price but in technical quality."

There is a great pool of managerial talent which could be unleashed for the creation of a better life, a higher standard of living, better schools, and a host of things which are scarce in this so-called age of abundance. Our response in time of war has shown the enormous reserves of idealism and commitment to service which lie locked in the American society. What a tragedy it would be if, by our failure to open communication in the discussion of these problems, or by the assumption of attitudes frozen in images of the past and polarized against change for the future, we should have our future pass behind our backs.

DISCUSSION

THE ACCELERATION OF TECHNOLOGICAL CHANGE

The premise of currently accelerating technological change must acknowledge past trends. For example, in the late 1920s, the A. O. Smith Company, a producer of automobile frames, displaced 99 percent of their labor through the use of machines. Very few recent developments have been so spectacular.

The statistics seem to show that the rate of growth of productivity between 1940 and 1955 was a bit faster than it had been before, but since 1955 it has been much slower. Although we have increased gross national product by 60 percent in the postwar period, only one-fifth of this increase was due to an increase in the number of man hours worked. This increase was actually largely due to increased efficiency.

Many of the problems of accelerated technological change derive from the present state of western culture, rather than from the advanced level of our technology. They are found in Europe as well as here. One serious problem that is emerging is that of chronic unemployment. And this leads to the related problem of convincing both educational and industrial institutions to take on the job of retraining and upgrading the labor force.

It seems to many that the important problem is to accelerate growth to a level near the rate of the Soviet Union. The lagging sectors of the U.S. economy are extremely large. We do not give much thought to improving the productivity of the Post Office, yet it employs almost as many people as the railroad industry.

Although it is difficult to prove that there is an acceleration of technological change, there are three indications of a discontinuity with past rates of technological change. First, there is the emergence of the machine that replaces the skill of man. We have, today, instead of the skill of man together with the power of the machine, the skill of the machine with the power of the machine. Second, there is the problem of technological unemployment. Third, there is the revolution of abundance. If we utilized freely all our new technologies it might be possible to make all necessities and conveniences free goods within ten years.

However, the question of whether technological change has accelerated should not be reduced to a question of decreases in input of man hours or increases in the output of goods and services. The population itself has a sense of the rate of change. People are more aware today of a gap between their folk knowledge of physical processes and the actual level of scientific advance. This produces a feeling in them of alienation, of losing out.

Certain breakthroughs, as in the area of space technology, represent an order of change which is qualitative in character. The continuities and discontinuities in this spectacular development are still to be shredded out. Many of the radical changes in technology have not been accompanied by adequate adjustment in our social institutions. In order to preserve our society we will have to consider the impact of technological changes on social structure and social performance, as well as how progress in scientific knowledge can be accelerated and transformed into further technological advances.

A further challenge is acceleration in the acceptance of technological change, not the acceleration of technological change itself. The role of the consumer with regard to technological change needs a great deal more study. A great deal of change is possible, not only in distribution, but in more fundamental social patterns. Our attitude toward work must change. These new technological changes have a totally new facet. They earn time. One of the great social problems now is what to do with that time.

THE CHANGING ROLE OF THE TECHNICIAN AND SCIENTIST

The pivotal force in the development of culture is now in the hands of the doers rather than the thinkers and the centers are now the great corporations rather than the universities. It is in these new centers that new directions are charted. The danger that the entire culture may become technological is obvious. Even in the recent past the intellectuals, the professional intellectuals, knew that they were the leaders of thought. But the doers are assuming this role and are doing a better job than the intellectuals.

In addition, there has been a great shift of power to people who are working on the frontiers of knowledge. An example of the influence of scientists came right after World War II, when nuclear physicists became the advisers to the groups in Washington who were making policy. At that time, the scientists were the only ones with access to very specialized information and they took on roles far outside the proper jurisdiction allowed by their scientific knowledge. A more complete shift of power toward the technologist is probably not inevitable, however, because of the latent strength of the American democratic tradition.

Because of their increasing role, there has been a growing emphasis on a broader training for engineers and scientists; on instructing them in the humanities and social sciences. The engineering profession is deeply concerned with the moral and ethical responsibilities of the engineer. It is difficult, however, to thus broaden the university curriculum because students today are rarely challenged by history. The new corporate technological centers are better able to stimulate young engineers because they are more alive to the functional challenges of contemporary society.

However, although the corporate technological centers can do a better job of training or retraining the technologists than the universities, they cannot be innovators. They train very well for the specific job that must be done, but when a man becomes a leader, he must put his job in a larger context. The capacity to do this comes out of experience and out of a continuous tradition of leadership.

There is a possibility of discontinuity here. When a scientist or technologist is intensely interested in solving a problem, he becomes blind to other factors, regardless of how broad his education. However, as he rises in the corporation, he reaches a point where the responsibility of his job transcends that of the particular research operation he is managing. To meet these larger responsibilities remains primarily a function of the human experience of learning through failures and successes.

Of course, the people who occupy leadership positions in management can hire technologists. But often they are not able to transfer to them the fabric of their values. It is becoming difficult to continue the apprenticeship of wisdom that normally flows between the man who manages and the technologists. The manager may feel that his outdated technical training is inadequate and this fear may be communicated in the transmission of his values as manager to those next in line. It is essential that the non-scientist stop worrying about the gap between the two cultures and admit that there is a management of science.

THE PROBLEM OF COMMUNICATION OF TECHNOLOGICAL INFORMATION

Knowledge has become the major instrumentality for accelerating change. Those who can manipulate the frontiers of knowledge become very powerful. But such people comprise a relatively small and esoteric group who cannot easily communicate with other people. In a modern democracy, although groups of scientists control specialized information, each citizen has a responsibility to share in the shaping of that society. The scientific community therefore must attempt to communicate esoteric information to the citizen in a form which will permit him to make reasonable judgments.

In addition, there is the more specific problem of communication between the technologists and management. However, managers can force the technologist to communicate with them. There is not a scientific problem that cannot be explained in a clear-cut manner to a person of intelligence if the technologist wants to explain himself. Leadership of this country, both in industry and in government, must insist upon this communication.

How can a technician explain technology to a non-technician, namely a manager? Part of the intellectual community must assume the role of translator. Perhaps we can develop a minority of technologists who are able and willing to translate. This might help to solve the communications problem.

THE CHARACTER OF LARGE INSTITUTIONS

The fact that organizations are becoming larger and simultaneously searching for stability lies at the heart of many of our problems. As the large organization gets larger it develops a definite personality. It tends to specialize functions and it becomes conservative. It loses the quality of risk taking. We must somehow create an environment in which our present organizations and institutions will be more innovative.

Another problem is that the supply of very bright people is nearing exhaustion. Organizations consequently have started to break up jobs so that people of little ability can do them. This leads to a very big problem of control over the direction of the corporation.

Let us consider when breakthroughs in ideas occur. Can we identify an economic incentive? The typical large business organization seems put together to produce "safe" decisions. People are likely to behave in a safe manner if they work for a long time in a large organization. Second, if we can identify organizations which produce risk takers, we might find a reservoir of potential or a method for increasing it.

Risk taking has assumed a new dimension today with the structure of the defense effort. This is a new phenomenon on a new qualitative scale. So much energy and inventiveness goes into these areas. How can we get some of that energy and inventiveness to spill over into traditional institutions? How much longer can so much of our innovation go into the defense area?

The key practical problem is to find a way to tap the resources locked up in organization structures. They do exist because periodically we get major product changes, although we may have to wait generations for an outstanding development. They are necessarily very infrequent because there is only a limited supply of very high ability.

The Post-Industrial Society

2   by Daniel Bell

Professor of Sociology, Columbia University

The post-industrial society is a society in which business is no longer the predominant element but one in which the intellectual is predominant. The majority of the society will not, of course, be intellectuals, but the sense of the society, its spirit, the areas of conflict, of advance, of engagement, will be largely in intellectual pursuits. The major institutions of society will be a vast new array of conglomerations of universities, research institutes, research corporations.

Three factors are important in giving a new character to society and a new character to its problems, assuming no basic change in the military situation and approximately the same kind of military establishment as now exists. These three elements are primarily responsible for the new character of American society as it is already visible and they will become increasingly important over the next thirty years. They are 1) the exponential growth of science, 2) the growth of the intellectual technology, and 3) the growth of research and development activities.

What are the dimensions of these three elements? Perhaps the most important is the extraordinary emergence of science and its increasing hold on the world around us. The last attempt to write a synoptic account of the world of science, Comte's *Philosophie Positive*, is a stupendous work. It is intriguing today however, to remember Comte's citation of the chemical composition of the stars and the existence of forms of life on them as "inherently unknowable." A few years later, spectrum analysis provided the very knowledge that Comte had thought to be impossible. Few people today would declare with confidence that something is unknowable. We assume that there are no inherent secrets. This marks a significant change in the way we approach problems—everything is open, nothing is unknowable. This is one of the hallmarks of modernity, a welcoming of change, an awareness of change, and a struggling effort to control the pace and direction of change.

But the recency of this development of science is rather startling. James Conant tells that in World War I, he, as president of the American Chemical Society, offered the services of the society to Newton D. Baker, the Secretary of War. He was asked to come back the next day, when he was told that these services would be unnecessary because the War Department already had a chemist. He also tells about the board, headed by Edison, which was created to aid the Navy. It had one physicist on it who was put there by Edison who said to Wilson, "We ought to have one mathematician fellow in case we have to calculate something out." Compared with the role of science in World War II these stories give a sense of the rapidity of the development of the role of science in military affairs and in the whole life of the society. Clearly there is something *new* when the dimension of science is enlarged so rapidly.

Other kinds of evidence indicate the growth rate of scientific knowledge. Some, like Gerald Holton, have expressed a degree of skepticism about the quality of this new knowledge, but it is the sheer quantity that is significant here. We know about the information revolution, the 50,000 technical journals which publish annually 1,200,000 articles. The chemical abstracts alone come to 13,000 pages a year. This growth of the output of scientific knowledge seems to be occurring at an exponential rate. The holdings of major libraries, according to an estimate by Ridenour, can be expected to double every sixteen years. This was in fact the experience of the Columbia University library over the past sixteen years.

It is not only the growth rate of the production of knowledge which is significant. There is another phenomenon, the proliferation of fields. The national roster of scientific and technical personnel already has 900 distinct categories. It is in the advanced areas of research that the most rapid rate of proliferation of fields is taking place and it is also in these areas that the most rapid rate of growth of knowledge is occurring.

This growth rate of scientific knowledge is an element which is currently remaking the universities, the research institutions, and the corporations that are involved.

The second *new* factor is the growth of intellectual technology. There is at present an effort under way to duplicate almost every kind of human skill through computers and through forms of theory independent of specific individuals. We are creating self-regulating systems. There is even an attempt to create a kind of Tableau Entière, to create concepts of rationality to cover all areas of knowledge, on the model of Quesnay's Tableau Economique. There has been an extraordinary rapidity in the development and spread of skills like decision theory, utility preference theory, operations research, game theory, cybernetics, and so forth.

The significant point is the onrush and application of computers which are able to perform a whole range of problems impossible before. There is even a presumption of the possibility of performing controlled experiments in the social sciences through simulation techniques. These developments have already played, and will play increasingly in the future, an extraordinary role. They have all sorts of ramifications and will contribute importantly to the coloration of specific features of the society which is emerging.

The third *new* factor, again taken on illustrative level, is the increase in the importance of research and development activities and some of their consequences. The growth rate of expenditures for research and development has been between 10 percent to 20 percent annually since 1920, depending upon which estimate is used. In dollar terms these expenditures have grown as follows:

1920	$80,000,000
1930	$130,000,000
1940	$377,000,000
1950	$2,870,000,000
1960	$14,000,000,000

We will not discuss here the technical problem of how much the published figures should be deflated to give true magnitudes. However, David Novack has estimated that the most recent figures should be deflated almost 50 percent, largely because of a change in the accounting procedures of the military establishment which used to put research and development expenditures under the general heading of procurement. Even with such deflation it is clear that in an overall perspective something *new* has happened to the society and the economy.

Equally clear is that the cost of research has become tremendous. The experiments by the Columbia University team to split the neutrino came to about $1,000,000. Compare this with expenditures in the creation of prototypes of missiles and other space technology. These have kited impressively the cost of research and development. The complexity of modern weaponry compared with that of even the recent past is illustrated by the Norden bombsight and the analog computer bombsight of the B52. The former cost $2,500 and could be carried by a man. The latter cost $250,000 and weighs between 1,000 and 2,000 pounds.

A large part of research and development is an outgrowth of the increasing complexity of weapons systems and the greater degree of precision that is required by them. The military are obviously going to continue to be the chief source of funds for much of this.

The experts disagree about the future rate of growth of research-and-development expenditures. Brozen, for example, feels that it will slow down somewhat in the next decade. Nevertheless, the new industries in this area already present new problems. The executives at Aerospace, for example, say that they face problems not of mass production but of mass research. Their problem is to administer several thousand people who are involved in different research projects.

Coordination of research is becoming crucial. Central to all of this is the role of the federal government. Wiesner claims that one third of the *free* funds of the federal budget currently go to research and development. The government is the chief source of research-and-development funds and the chief beneficiaries are universities and corporations.

This situation has produced a new problem in the relations between the government and the universities and corporations. Formerly, the federal government contracted for mass-produced items such as ships, planes, and tanks. However, when the federal government deals with universities and corporations today, it is dealing with research ideas or with development. This presents a very different problem in terms of degree of control and organization.

A few of the problems and social consequences which emerge from this new situation are: 1) the crucial role of government. The cost of research, using cost in a wide sense, can only be met by government. The total costs of change are so great today that very few companies or even industries can afford them. The cost of technological change in the longshore industry is being handled by the whole industry, rather than individual firms, but this is a small industry. The cost of social change in the coal industry is being borne to a large extent by the miners themselves in the form of chronic unemployment; the federal government will inevitably have to shoulder an increasing proportion of such costs. 2) Insofar as the basic drives and energies of society go into research, the degree of control exercised by either government bureaucracy or scientific bureaucracies raises another series of critical questions. 3) The problem of the formation of "human capital," a problem which is engaging a number of economists, becomes more acute. If a society is primarily geared to research, the cost of the waste of "human capital" becomes increasingly important. The raising of "human capital" is a much more complex process than the raising of financial capital. It is the limits to "human capital" rather than to financial capital which have become the fundamental element limiting the growth of the society. It is not only the raising of "human capital" which is important. The problem of deciding what kind of "human capital" we should have becomes increasingly important.

There is another way in which changes in technology are likely to have broad human and social implications. When a group in society finds itself dispossessed, a degree of rancor sets in. As technical competence becomes increasingly the criterion for a position in society, for advancement, for mobility, an increasing number of specific groups may become dispossessed. This of course will create social tension. The growth of the radical right among certain elements in the military establishment and in business life in America may be a product of this kind of dispossession. The intellectual society of the future is not going to be a neatly arrayed scheme wherein the hierophants of science rule in a very calm way, as Wells predicted. Rather, it will be a society in which strains may increase and the political counterparts of these strains become more rancorous.

The greater emphasis placed on educational competence will mean that groups which lose out early in the educational race will be quickly excluded from society as a whole. In the next twenty or thirty years the economic situation of the Negro may become relatively worse, simply because the rate of economic change is such as to outrun the increase in educational opportunities available to him. Apart from a thin stratum who do have better educational opportunities, a large part of the Negro population, more than half, continues to live in the South, many in the agrarian sector. This agrarian population has been relatively excluded from society. In this sense coming into the industrial workforce was a way for the Negro to enter modern society. The fact that a large proportion of the Negro population continues to live in this agrarian situation, many of them functionally illiterate, some of whom will emigrate to the North, means that the position of the Negro will worsen. A disturbing indication today is the dropout rate for Negroes in schools in New York and Detroit. This rate is an indicator of the position of the Negro thirty years from now.

Lastly, the growth of technical specialization creates a strain on the cultural level of society, the term used here not in its anthropological sense but in the sense that culture is the symbolic expression of what is occurring in society. People try to symbolize their experiences in order to make them intelligible to each other. This is the way nineteenth-century culture developed, a culture that arose out of an awareness of social mobility. The novels of the time represented the way in which the awareness of the new experiences of society found symbolic expression. It will become more and more difficult to find common symbolical expressions of the forms of specialization that are now developing. As a consequence an increasing disjunction between the culture and the society may arise. This is not just a problem of the political alienation of the intelligentsia but a much more pervasive problem of the inability of the society to find cultural terms for expressing what is occurring in the realm of science and in life itself. This is not just the "two cultures" problem presented by Snow because it is not just a matter of the education of people. It is, rather, the problem of the inability to find symbolic expressions for the kinds of experiences that take place in the work life created by the new forms of intellectual technology.

DISCUSSION

THE "NEW" ELEMENTS IN SOCIETY

Saint-Simon foresaw an industrial society as follows: The wealth of such a society is created by production and machinery rather than through the old methods of war and exploitation. The society is organized by the industrialists in a positivist fashion, using a methodology of order and precision rather than a metaphysical order. And the society is structured in accordance with the functions of individuals.

Then came Comte, one of the last of a series in the great tradition of confidence in reason and in the capability of rational scientific organization to solve all scientific problems. The striking point today is the demise of that point of view. Twentieth-century thought is characterized by the lack of confidence in the rational solution of social and cultural problems. This lack of confidence is particularly striking when we think of the range of scientific endeavor under way today.

The basic premise of our economy has been that the decisions of the businessman would work out for the common good. This is no longer true. There are three "new" problems: 1) human capital, which cannot be handled on the same basis as financial capital, 2) the information revolution which will bring our society to a grinding halt unless it is solved, 3) differences in the pace at which various parts of industry and society are moving. We need an entirely new set of conceptual tools for the solution of these problems.

It was stated in the last chapter that there is a direct correlation between technological acceleration and economic growth. However, technology is but one of the factors making for economic growth.

The future areas of real expansion are the service sectors. Agricultural employment is constantly decreasing and there has been no increase in manufacturing employment in ten years. Jobs will probably continue to decline in agriculture and manufacturing, perhaps also in construction and materials handling. But there are three fields which are expanding rapidly and there seems to be no limit to any of these areas. They are recreation, education, and health.

What are the kinds of institutional and structural problems we face in trying to get the kinds of services we want? What kinds of enterprise structures must we create in order to attract the kind of manpower we need? The American public today, a metropolitan public by and large, requires a series of services which American private enterprise cannot produce because of barriers to investment and to sale for profit. Therefore, the public cannot secure what it most wants and needs and could afford to pay for. There seem to be major blocks to the exchange of services among the private and public markets.

Another basic problem today is how to couple the new science and technology with practical applications which will actually result in economic growth. For example, Japan spends 0.9 percent of its gross national product on research and development, largely in the civilian sector, and its rate of economic growth is about 7 percent. We spend almost 3.0 percent on research and development of which only about 10 percent is devoted to the civilian sector, and our growth rate is lagging. That means that for the entire civilian sector we spend only about 0.3 percent of our gross national product on research. That is why this sector, which represents about 70 percent of our economy, does not grow.

Although our economic growth rate may have leveled off, the growth of science continues. Professor Derek Price of Yale University has written an essay on the exponential growth of science in which he points out that if the rate of growth continues, there will soon be more scientists than people! Perhaps we have an exponential growth of scientists, not necessarily of science. There is an analogy which likens most of our journals to a squid that moves backward ejecting ink.

Most of the growth of science today, which is in the military sector, is not easily adapted to the civilian sector. Both the chemistry and communications industries first grew substantially after our government became involved in them during World War I. The National Academy of Science was started after the Civil War to handle problems faced by the Navy.

There is very little spillover from industries involved in space development into the civilian sector of the economy. What the government spends on military or space devices rarely makes a direct or even very slow indirect contribution to the civilian sector. The whole development in these new areas is different. Distinctions must be made between the new technology and the problems of dovetailing it into the different parts of the economy.

Nevertheless some authorities have claimed that we have a kind of scientific capital which is being used up in its application. As long as scientific development is tied to utilitarian ends, we might be using it up faster than we are replacing it. Professor Margenau of Yale has pointed out the decreasing time interval between pure scientific advance and its application.

Most of the discussion so far has been carried on in terms of the American economy. However, we comprise only 6 percent of the world's population. If we set our scientists to solving the problems of the rest of the world, we would not have to consider the problem of abundance, but of scarcity, particularly if the world's population continues to double every fifty years.

THE FRAGMENTATION OF SOCIETY

With reference to power in society, there is a base of power and a mode of access to power. In the past, the base of power was property; the mode of access, inheritance or certain forms of entrepreneurship. There is a second base of power in the modern mass society, which is mass voting; the mode of access is the mobilization of voters through the modern political machine. Recently there has come to be a third base of power, skill, and the mode of access to this power is education. At any given time all three modes of power exist and it is difficult to be precise about the exact dimensions of each mode. This is a process and any process presents this difficulty.

One of the central problems of the future is the fragmentation of society. This will not be just an economic problem. Those who lack understanding of the new society and do not have a sense of involvement are going to be dispossessed. We should be interested not just in what happens to the economy but in what happens to the people who have a tradition of how a democratic society should work and feel that somehow it is failing. It is not clear that providing people with a scientific education will be a sufficient remedy for this kind of problem.

The new technology changes the worker from machine operator to machine overseer. This is possible only because of very large prior expenditures on the initial installation, but once this is accomplished, the operation does not require educated personnel.

We can picture a society which is becoming increasingly depersonalized. This will make institutions like trade unions increasingly alienated. This is one of the basic problems which will arise from accelerated technology in the future society. The answers to this problem will not come from existing institutions such as the trade union movement. The responsibility must devolve upon the government. None of us wants to live in such a depersonalized society. It is to our mutual interest to see that the kind of institutions we need for a democratic society will get the kind of support that they must have in order to play their necessary roles in this future society.

THE MANAGEMENT OF SCIENTIFIC RESEARCH

The most important institutions of society used to be the business firms. They were run according to certain rules and constraints, and the principal constraint was that of competition. But other kinds of institutions, some of them governmental —but, even more important, those that are neither governmental nor private, ranging from the mutual insurance company to government authorities such as TVA or AEC and, in between, the universities and research institutions—will increasingly contain the basic energies of society. These institutions certainly do not run to the same rules as the business firm and so far economists have had very little to say about efficiency criteria for an economic system in which these kinds of institutions are very important. The management function today is the coordination of human beings, not simply that of increasing productivity. One of the management problems to come will be that of increasingly managing scientific research. Research and development is as well managed today as American corporations in general are managed today. Nevertheless, research is not easy to manage. It may become so institutionalized that it will send many people down the same paths while a lot of important problems are neglected.

Several questions arise. How much of the scientific work of the future will be creative? At any given time scientists are so bound by the presuppositions of their scientific community that it is extremely difficult for them to move beyond them to a real scientific revolution. And if an increasing weight is attached to the existing structure of scientific knowledge, this may make it even more difficult for scientific work of real originality to be accomplished.

An indication of this is that the number of patents granted in 1960 was almost the same as in 1930. For example, all the papers that were written on ether at the turn of the century came to nothing. There were a few brilliant papers and they were what really counted.

Can we insure a proper balance of activity? Has initiative remained where it was or has it gone to the government? If, in the new society, there will be new institutions which will take the place of the business corporation, such as research laboratories and universities, who will control them? If the state is to control them, can we consider the state an intellectual institution?

Uninterrupted linear development of science and technology is not the ordinary process. Moreover, this does not occur in a political vacuum. Given the American political environment, the process inevitably becomes meshed into a political process which ultimately would tame it, subdue it, manage it. Sheer uncontrolled development is inconceivable in our kind of society.

One of the striking aspects of government's problem is it has not been able to call in businessmen for the solutions, but has had to call on such institutions as RAND and MITRE. This has led to the extraordinary proliferation of these institutions and represents radical departures in the way our government and our society operate.

HUMAN RESOURCES

The educational requirements for employment in industry have been increased. One of the reasons lying behind much of the large expenditures for research and development is that this is a method of educating people to higher levels. The increase in funds proposed for the National Science Foundation is directly related to this: the Foundation is becoming a mechanism for raising the educational level of scientists.

However, although we always need more "preferred" people, more educated people, more healthy people, more beautiful women, there is often a discrepancy between what we think we need and the actual shortage. For example, in the most rapidly growing sector of the economy, the medical sector, there is a discrepancy between the educational ambitions of nurses and the actual needs of hospitals. The administration at the Menninger Foundation has concluded that in mental hospital attendants, qualities of character and personality, rather than formal education, are crucial. The presumption that we actually need more scientists, technologists, and technicians may be contrary to fact and has no clear relation to the performance of many functions.

In connection with the increased demand for more education for more jobs, the distribution of income is still a problem. There are still large numbers of people with very low incomes. From a social point of view and from the point of view of economic efficiency in terms of the development of children in such families, their health and education, this is a very costly situation and we can afford to remedy it. Unfortunately, however, we do not have any clear ideas about how to do so within the framework of our present economic system. The handicaps of those with little education are going to increase largely as the result of scientific change and technological development. The solution of this problem will also have to be based on a rethinking of some of our preconceived doctrines.

SCIENCE AND CULTURE

The rise of modern science has had a tremendous impact on contemporary music and art, but the autonomy of artistic activities is fairly general throughout the history of Western culture. Nevertheless, the effect of science and technology on culture has been substantial. The effect of science, for example, on poetry has been to drive it out of the area of fact and into the area of ambiguity and ellipsis, simply because fact is handled so much better by science.

There is at work today a much deeper cultural process which is having extraordinary consequences for society. A classic illustration of the union of science and art was provided in the early Renaissance by Uccello. He emphasized in his paintings a mathematical specificity of perspective in order to get a vision of reality. Painting to him was an attempt to create an illusion of three dimensions in two through the use of mathematical techniques of perspective. This is an example of the height of a rational cosmology, a cosmology built on the ideas of beginning, middle, and end, foreground and background, the Aristotelian unities in the drama, and the total rationalization of perspective in painting.

In our culture today we see the end of this rational cosmology. In a very simplified sense, there is an end of linearity and the emergence of the problem of the creation of simultaneity. People no longer can have a sense of linearity, of beginning, middle, and end, foreground and background. One can see this for example in the novels of Faulkner, in serial music, and so forth. People respond without being entirely conscious of the multiplicity of interactions, the multiplicity of experiences. The disjunction between science and culture is not merely that people cannot find words to express experiences to others. The whole breakdown of the rational cosmology is imminent and will ultimately create the most serious problems for society as a whole because of an alienation of the modes of perception about the world.

The Aerospace Industry

3   by Earl D. Johnson

*Executive Vice-President, Delta Airlines
and formerly Vice-Chairman of the Board,
General Dynamics Corporation*

This chapter will discuss some of the problems of a major contractor in the aerospace industry, particularly problems involving the use of technical manpower. How, for example, should the contractor adjust his manpower requirements, particularly for scientists and engineers, to the changing nature of the defense business? Or again, how does doing business with government differ from doing business with the private sectors? To what degree can the lessons learned from government work spill over into the civilian sectors? What developments that we see occurring in this fast-moving area give some indication of future developments in the total economic and social structure?

We will first address our attention to the question of the difference between the scientists and the managers and the possibility entertained by some that management might gradually be replaced in the decision-making process by the scientist and the engineer. There is no question but that in the aerospace industry the scientist is being heard in the councils of business, but this is not by any means a replacement of management. It represents, rather, an addition to management. The scientist and engineer are merely additional voices in management.

In this industry the scientist does not merely address himself to the technological phase of the business. In order to sell programs in competition with other companies, the scientist becomes a salesman, just as the salesman becomes an engineer or scientist. The scientist-salesman is used in all facets of the business. Nor is there concern on the part of business that the manager is being replaced by the scientist. There comes a time in the economic development of a country or of a particular industry when different people with different skills are required to run the affairs. Certainly during the 1930s we needed people with top accounting and top financial skills as the managers and this is what we got. At present in the aerospace industry, the problem is that new products large enough to sustain the organization come along so infrequently that the scientist or engineer who comes forward with new ideas, who at the same time must translate these ideas into a product, can dictate where the money should be spent. It is, however, unlikely that he will continue to determine the channeling of the funds if the industry is forced to retrench.

In every aerospace corporation, there has been a gradual diminution of the number of production workers proportionate to the number of scientists and engineers. There has been a shift from the repetitive type of production work to a greater and greater emphasis on the individual's contribution to the development of an individual product, which usually turns out to be a one-time product. For example, in World War II this country produced as many as 90,000 airplanes in one year. By the time of the Korea conflict, peak production had dropped to 10,000 a year.

Today when we talk of missiles, we are talking of hundreds and of those hundreds only a very few are identical. This means that production becomes a matter of individual design, prototype, some flight testing, and then a gradual growth into an entirely different missile.

All of this requires men of great technical ability, of great scientific skill. It requires a very broad spectrum of people from almost all the disciplines. For example, General Dynamics has one division of some 1,300 people of whom over 500 are engineers and scientists. This division was started in 1955 with a man who was thirty-two years old as the head. He was given a desk and a secretary and told to start a division. Naturally he hired brilliant young people. These people were in their twenties in 1955 and now they are in their thirties and some of them are in the forties. A problem has emerged in dealing with these people. As they get older their economic requirements increase, but according to the studies we have seen, their scientific productivity in terms of ingenious ideas may fall off. But in order to hire these men in the face of the competition of other firms, the company had to pay them initial salaries that were commensurate with salaries ordinarily paid an age group of forty or forty-five. Now the company is faced with the problem of whether these men have dropped in value. Should management go ahead and increase their salaries and give them other responsibilities? Unfortunately, as one moves up the corporate or any organizational pyramid, there is less and less room at the top. Consequently, these able young people represent a major problem, even for a young company.

The large aerospace contractor does the overwhelming bulk of his business with the government. It is important that we have a general conception of the overall dimensions of defense spending in the United States today. The U.S. defense budget of $56 billion has shifted from mass production to custom-built technology. Moreover, not only is defense contracting the biggest industry in the world, but the most complex. And in addition to its complexity, it is unstable.

It is complex because it produces products which are as complicated as bombers with 100,000 electronic components apiece and space capsules that require the coordinated efforts of nine thousand separate companies. It is unstable because the swift pace of technology makes it so. A weapon that takes ten years to design can be obsolete in five, and the company that climbs to the ranks of the top one hundred defense contractors finds itself in a group whose membership is twice as unstable as that of the top one hundred industrial firms. It seems that the large defense contractor goes up and down like a yo-yo compared with the rest of business!

A further comment on the custom-built character of defense production is in order here. Deputy Secretary of Defense Gilpatrick has said that much of our future production will be custom-built equipment based on the latest research rather than on mass production. The technology is moving with such rapidity that we can now pass through a whole weapons generation faster than some can be developed. For example, the generation of air-breathing missiles, a tremendous technical breakthrough, was obsolete and discarded before any of them ever entered the inventory. Once a missile gets into the air, a 50-cent switch can knock it out of operation though it may have cost $10,000,000.

As a result there is an emphasis on reliability which never existed before in industry. The technology is unbelievably difficult and the reliability standards almost inconceivable. Defense contracting is so vulnerable to rapid obsolescence that, as recently reported: "More than 52 major missile programs have been canceled at a cost of over $6 billion." Regardless of possible shortages, then, it is clear that the aerospace industry is a hazardous area for financial capital. This is the sector of American economic life which is most affected by technology. Moreover, until the recent past the aerospace industry has employed more people than the automobile industry.

Since government is the prime customer for this industry, with all that implies in terms of organizational structure, personnel selection, training, the construction and acquisition of facilities, political considerations must be faced even after production has been made as technologically correct as possible. For instance, there is a law that government must buy at the lowest bid. An example drawn from the Korean War illustrates the kind of problem that sometimes arises. The Army had to buy soap from a Brooklyn manufacturer, who submitted the lowest bid, even though the total cost of the soap delivered in Korea would have been less if the Army could have bought it from a West Coast plant.

A different set of considerations develops out of the small business administration's activities which force certain adjustments. The existence of depressed areas such as Wilkes Barre and Scranton raises other considerations in the actual location of defense work. In other cases the Defense Department may feel that they must place a contract with another company whose bid is higher and whose technology is second best, because it is essential to have a second source of supply. These political and military considerations increase the degree of instability of defense contracting. Unfortunately, the effects are not confined to stockholders; they apply to executives, workers, and everybody else.

It is important to arrive at a better definition of research and development expenditures. In the Army it can be the theoretical conception, it can be the design, it can be the testing, it can be the prototype. To complicate the concept: if research and development has been denied sufficient funds it is possible to spill over some of the actual production money into research and development. The fiscal 1963 defense budget calls for about $5 billion for research and development. But informed people agree that of the $5 billion not over $100 million will be for basic research. Similarly, individual companies are unable to break out what they spend on their own research and development. It is impossible to identify clearly such items from a company's books.

Until more is known about the actual composition of such expenditures, we seem to be trying to draw conclusions from insufficient facts or facts that are not clearly understood. Government has one definition, business another, educators a third, and the scientists have their own. Moreover, government research and development is unique. It involves areas of technology that are not needed in industry. Industry will use some of these developments, but they really do not need them.

The difference between government research and development and civilian research and development springs from the fact that the military must always push for the ultimate in performance by the individual and by materials. This ultimate can be achieved only by pushing forward on all the technological and scientific fronts simultaneously. We could not have produced the atomic bomb unless a whole spectrum of sciences had pushed forward. No individual business can afford this.

Only the government has sufficient resources to tackle problems on this scale. The technical fronts be pushed forward simultaneously; the military must try to telescope time. As a result the theory, the conceptual design, the prototype, the testing, and even some of the production goes on concurrently. Of course this makes for astronomical costs. Duplication is necessary because the directors of these projects cannot be sure of which effort will come up with a successful result. This accounts for the fact that 52 missile programs were started and canceled in recent years. The problem is getting worse because the weapons systems are becoming much more complex and the time scale more and more difficult to compress. The Skybolt and much of the development of the Dynasoar and Gemini programs are illustrations of these facts.

These are some of the underlying conditions which face the large aerospace contractor when he deals with the government.

Compounding the problem of employing high level scientific and engineering personnel is the fact that when the government cancels major projects, no company can afford to keep these people on the payroll while waiting for the next program. What is management to do with these people between major projects? Since they cannot all be kept, a priority must be established and the rest let go. But, for example, some of the people thus necessarily dismissed played a major role in the development of the supersonic bomber upon which the defense of the free world rests. Now they have no job and cannot go to similar companies because they are cutting back too. The people who bear the brunt of this instability are not the traditional production workers who were looked on as a commodity in the 1920s. These men are brilliant scientists, graduates of the finest educational institutions in the world. They will not stand for this kind of treatment. Some additional thought about human relations must be applied to this area.

The problem of management in relation to technology is also compounded by the enormous importance of government expenditures in the area of research and development. Is it ever going to be possible again for business to recapture the leadership from government in research and development? Probably not. Moreover, the leadership which originally was based largely on military needs is in the process of shifting its interest and orientation toward projects such as NASA. Almost $20 billion is now being scheduled for NASA through 1968. In one sense these expenditures are almost entirely research and development. It appears then that government will permanently be the leader in this area, and will therefore be the biggest source of employment funds—of the scientist, the engineer, and the technologist of the future.

Now it is only a short step before government administrators of these enormous programs say that as long as they are the biggest employer, they had better have management of these funds. Indeed, the prerogatives of operating management are being encroached upon daily. If the country is short of scientific and technical personnel, we had better employ them more efficiently. But this implicitly questions important elements of the capitalistic system, such as the right of a man to work at what he wants, for whom he wants, and how he wants. If there is insufficient technical manpower to maintain the supremacy of the United States and we are using it wastefully and the government itself is becoming the biggest factor in research and development, this may force us to move toward increasing state control.

In preceding chapters the question of the time lag between the discovery of scientific principles and their practical application has been discussed. There is an interesting statement of this: "Photographic principles were set forth by Leonardo da Vinci some five hundred years ago. Yet the first workable process was not developed until 1839. The electric motor was developed by Faraday in the 1830s, but electric motor power did not enter the industrial scene until the 1870s. Radio broadcasting was based on the work of Maxwell and Hertz between 1873 and 1889, but it did not become a reality until 1904 and was not taken up by industry until the 1920s. Now compare these time lags between discovery and the use of a new process or product with those of today. Development of atomic technology, first through the Manhattan Project and then through the AEC, brought massive results in about five years. The transistor is one of the outstanding scientific discoveries of the era, yet here, too, only five years elapsed between its discovery and its widespread application in electronic computers, telephonic switching and radar. Wartime emergency and vast government financing played a significant role in each of those fields, reminding us that government participation, principally in the area of applied science and technology, has become an accepted stimulant to our society."

In addition to the fiscal forces and urgencies of war, another influence of perhaps equal effect on our economy is the influence of the new and suddenly appearing tools of science. Of these perhaps the most significant are the time shortening, almost time obliterating, electronic computers, and high-speed electronic devices. By means of these high-speed electronic computers, the discovery element, development, application cycles that lie at the heart of all technological progress are virtually telescoped. In many cases the lag between research and development fades out and all but disappears.

Often in any one company serious meetings are held on a project, a task force is set up, a key man is assigned to lead, even orders placed for some lead-time facilities. Then the project is cancelled. Why? Not because another company made the project obsolete, but because somebody in the same company made it obsolete.

With the increase in the speed of development of technology, education is no longer just the concern of the young. Education is a lifetime project and business, particularly the big business corporation, has come to realize it. Corporations know that they have to provide a certain amount of money for a certain number of people to be educated all the time. This is not just a question of training skilled and semi-skilled workers. It extends to the management schools and it includes all areas of a company's affairs. This expansion of the educational sights of business will permit businessmen to get together to discuss things other than the profit motive. It will allow them to do more thinking on the human problems which are arising and the government structures that are evolving out of this new situation so that solutions will begin to emerge.

A final point has reference to the uneducated. It is true, of course, that society is always going to be stratified. But at the same time education and the spread of technology is increasing the number of strata in the pyramid and perhaps widening the spectrum of the strata. Basically, everybody is moving up. The benefits of our society are spreading. There are no peons left. Of course, the people at the bottom are still at a disadvantage, but everybody is coming up. The good things of our society are more and more being shared by everybody.

With the spread of education, the country is becoming increasingly liberal. And as the country becomes more liberal it will be less willing to stand for technological unemployment and for chronic economic disaster. The country is going to insist that solutions be found for these problems.

DISCUSSION

TECHNOLOGICAL CHANGE AND ECONOMIC GROWTH

We are all concerned about the impact of technological change on our capitalistic system and on democracy itself. The greatest single impetus to economic growth is technological change which gets translated into goods and services—new goods and services and cheaper goods and services. However, today's tremendous research and development budgets take a great deal of money out of the private or civilian sector of the economy. Perhaps more important is that a great number of people are being taken out of the civilian sector. As a result of both of these diversions there is considerably less economic growth than there otherwise would be.

Without economic growth it is difficult to see how we will be able to finance such things as putting men on the moon.

These feats may be essential, but there is a great deal of difference between national defense and national security. There is not much argument about national defense. It must be secured. But national security is a much more complicated subject. One of its components is the economic structure of the country which is made of a number of distinct sectors. We cannot weaken any part of that structure without affecting the entire structure, including national defense.

How then can technological change be used to increase economic growth, particularly in those areas where there are social needs, such as urban renewal, transportation, housing, education?

DuPont, for example, has chosen as its primary mission to operate in the civilian sector of the economy. They believe that if this sector does not grow, the government will have considerable difficulty in financing some of its major projects for national defense. But they have decided that if they want to continue to operate in the fashion to which they have become accustomed and which they think is sound, they will not be able to work for the government and simultaneously operate efficiently in the private sector. Consequently, they have decided to stay out of the government sector, except in those instances where they are asked to do a specific job or where they can make a unique contribution.

Economic instability and consequent employment insecurity and displacement are not just the result of the displacement of men by machines. We have overemphasized automation. The effects of automation are not different from the effects of other changes which are produced by government procurement policy, for instance. Since it is the acceleration of change and the ability of people to adjust to change which is important, the crucial consideration is the nature of their skills and whether the adjustments take place in an economic environment which is growing stagnant or declining.

One illustration thereof, is the difference between the fate of two occupations in the railroad industry, the firemen and the boilermakers, both of which had their own unions. Dieselization hit both crafts very hard, the boilermakers proportionately harder. However, the firemen have a skill which is not transferable and a union which is limited to the railroad industry. The boilermakers, on the other hand, have transferable skills and a union which is involved in other industries. There has been no disturbance in the railroad industry over the position of the boilermakers, even though their numbers have declined from about 36,000 to 6,000, while at the same time the displacement of firemen has threatened to result in widespread strike activity. If the change is to be absorbed, we must have a labor force with transferable skills and an economy that is growing. We are going to have change, a lot of which we will never be able to anticipate, but we must concentrate on the nature of labor skills and upon achieving an expanding economy so that change can be made with a minimum of difficulty.

This is an age of specialization. To use examples drawn from paleontology and evolution: some animals, like the bear, have existed for a long time with very little change because they did not specialize. They could survive in any environment; in winter the bear went to sleep and waited for it to pass. But the koala in Australia eats only one thing, eucalyptus leaves. If a blight were to destroy the eucalyptus trees, the koala would be through. He is a specialist.

We produce many specialists and we pay them high salaries, but if the environment changes, an entire group may disappear. The difference between revolution and evolution, both of which are change, is the rate of change. We are approaching rates of change in certain areas that are virtually revolutionary and in revolution there are casualties.

The large corporation is a vast enterprise which lives on a day-to-day basis. It can quickly be renegotiated out of business or suffer vast losses, as Boeing did. A corporation in the field of governmental contracting is not at all like the old type corporation. Here the corporation seems to be a major part of the social structure of our society. At the same time the decision-making process is not a corporate decision any longer but is deeply involved in political considerations.

The importance of a major contractor in a specific locality can be almost overwhelming. A major cutback of a program then becomes a local disaster because every economic activity in the locality is so dependent on employment by the contractor. Considerations such as these were apparently behind DuPont's decision not to get too deeply involved in the defense business. It is also true that a defense contractor cannot make decisions on the same basis as an ordinary commercial company. This has been one of the difficulties which has come up when a defense company has moved into the commercial sector of the economy. In effect the defense contractor becomes a partner of the government.

It seems that human ingenuity ought to be able to ease the severity of the problem. For instance contracts might provide that when a defense contractor wins a contract he must subcontract 50 percent of it to the major contractor who loses. We could thus devise some means of easing the burden on those who happen to lose, assuming of course that all contractors who are able to bid for prime contracts are highly competent.

MANAGEMENT OF SCIENTIFIC RESEARCH

We have noted that management is still administering the engineers and scientists. But in every organization with an advanced technology the manager understands less and less about what is going on in the middle. As technologies change he never catches up. How many managers are there who are capable of keeping up with technological change? How many managements know what kinds of decisions they are delegating to their engineers and scientists—when they have made the decision to delegate at all? Many men in the upper levels of management do not really know the direction of research, and have not learned how to get information which will help them decide where the research ought to go.

Let us consider the various routes by which a manager arrives where he is. Certainly in some companies top management knows a great deal about their technology. But in the giant corporations today management can become very far removed from the actual day-to-day work. As the research operation becomes complicated, more and more managers have to come up the scientific route. They of course add their voice to management. Management becomes a composite. One of the reasons for the success of General Motors was the voice of Kettering, a top engineer and scientist, at the management table right next to the financial and legal experts. Management today must be a multi-faceted group.

The problem of management is a problem in every large organization, whether political, religious, military, or economic. It is not a unique problem that has emerged in the scientific arena today. Using DuPont again as an example: this company is governed by nine men at the top, five of whom are Ph.D's and all but one technically trained. They understand thoroughly what goes on in the company in terms of the products and in terms of the technology.

The management of scientific research is tied in with the problems of the impersonal relationship between the top decision makers and the people who are affected by their decisions. We all feel that it is wrong when individuals are affected by forces over which they have no control; and yet society would not be where it is today were it not for certain mechanisms which cold-bloodedly shield the generals or the President from all the human implications of their acts. Can we mobilize technology in order to get the best of both worlds? Can we have the kind of coordinated decision making that derives from having a man isolated on top and yet taking into account personal feelings and problems?

The question of managing knowledge in addition to men has been raised. What happened, for instance, to the knowledge that was garnered in the 52 missile projects that were terminated? Some companies have the problem of converting knowledge which has been developed on the frontiers through governmental sponsorship into the industrial area. Occasionally companies may hire consultants to advise them with regard to their nongovernmental projects only to find that the information that they are buying was developed on a government-sponsored project of their own company. The question then, is how to make newly acquired and important knowledge available to those who can use and need it.

THE ROLE OF GOVERNMENT

Government has been paying for two-thirds of all research since 1945. Industry may never be able to recapture the lead in research. And as a result some new form of social fabric and institutional structure is inevitably going to emerge.

How much difference does it make whether government spends tremendous sums for made-to-order weaponry or tremendous sums in other sectors? What effect will this difference have on the flow of funds throughout the economy? The made-to-order weaponry uses higher and higher grades of talent and less and less of the mass of the population. We know that it does matter how government spends money. In this instance it might make a lot of difference because of the nature of the circular flow. As a footnote to the above: What happened when we moved toward hard missile bases compared with the time when we were constantly enlarging and improving the inventory of weapons? Weapons became obsolescent very rapidly. The hard missile bases are not supposed to become obsolescent as rapidly. Does this have any implications for future output and employment? We are familiar with the effect of inventory cycles in the civilian sector. What about inventory cycles in the military sector?

If there has been so little effect on economic development from the vast military expenditures and if in the future there is to be a reduction in military inventories due to hard missile bases, perhaps the curve of expenditures on research may not continue to rise. For instance, the National Institutes of Health may already have reached the point where a levelling off is in order. We may already have the beginnings of a social reassessment of the flow of funds and the flow of people into research.

We seem to have reached a point where we have collapsed the time between discovery and application almost to the vanishing point. How much further can it be collapsed? As we have stated, the growth of science has been following an exponential curve over the last fifteen or twenty years. Inevitably the turning point cannot be far off. Is it these turning points which present the problem?

The question of the interrelations of the civilian and governmental sectors through military contracting has taken on new dimensions and is a new phenomenon in our society. We have a concept of markets and competition and other types of control mechanisms in the private sector of the economy. Now we have a whole new set of control mechanisms at work. Important political issues are involved. The allocation of contracts has effects on the long-term research potential of the country. What does it mean for a democracy when former high-level military personnel are hired to run companies which are involved in large scale business with the government?

SCIENTIFIC CAREERS

What does it mean in an advanced society when scientists and engineers can have no meaningful careers unless they are tied up in military technology? This is a new problem. In the past people of comparable ability and training—for instance, doctors—had a very secure and solid career. Now this part of our manpower—the trained scientist—is the most vulnerable, the most unstable, the most unsettled. With Hitler's Germany in mind, we know that it is not healthy for society to treat its brightest people so shabbily. The price can be very high. The question of social planning and control of the brainpower of the country in a more meaningful and less wasteful manner than in the recent past is a crucial one. This problem will become more acute if the expansion of the defense sector slows down.

In addition, the changing age distribution of our population means that, unless we find some new activities that will be respected and appreciated, many individuals whose contributions were very great when they were relatively young will find no productive activity open to them when they near sixty.

We cannot settle these problems or the problem of career development unless we have some idea of what the cross-section of activities open to a man at the age of, say, sixty should be. What should a nuclear physicist do when he can no longer be a productive scientist? This will increasingly be a problem since more and more of the areas of science and technology will be the provinces of young men.

This enhances the new necessity of continuing education as a lifetime proposition.

How do the scientist and engineer respond to the disruption of their careers and how does management handle these problems? One example is provided by General Dynamics which established a number of spectacular firsts in weaponry. As a result the inquisitive mind and highly intellectual scientist wanted to go into these projects in order to be associated with the new. When its own programs were coming to a termination, the company was able to siphon off a number of people to the programs of other major contractors. There were enough new programs which sprang out of the development of NASA to facilitate the transition.

But the problem of the future may be much more difficult. The shock of having major programs run out is not easily absorbed. Some of the space programs will probably be cancelled and this will make the absorption of some of these scientists difficult. Moreover, some of the brilliant men who were involved in these programs will not have the flexibility and productivity that they had when they were in their twenties or thirties.

Second, it may be very difficult to use scientists and engineers who have specialized in one area in other areas. The example of GM engineers in the Navy comes to mind; the Navy found it hard to use these people at all.

There seems to be an analogy between the boilermakers with their transferable skills and the highly skilled, but highly specialized technologists. One of the problems here is that the scientist feels that he can specialize to a high degree, that he can spend several years working in a very narrow field, and then find a company which can use his more general scientific skills. In actual fact, very few companies in the civilian sector of the economy can use the kinds of highly specialized skills that are being developed in the military and space sectors.

We need much more flexible forms of organization than government or industry now have. We need sound social mechanisms which give people both security and flexibility so they do not have to worry about being laid off every six months or a year.

The problem of sequential careers is coming up. Just as in baseball where players become coaches and eventually open restaurants, so perhaps there will be movements from private industry to government, then into overseas jobs in much greater numbers, and back again into a teaching circuit. But up to now we have not done much to facilitate this process. The increasing rate of expenditures on research and development made it possible to absorb these people easily in the past, but there may be real trouble in the late 1960s. Many of the universities themselves are dependent on the maintenance and increase in the level of governmental expenditures of this kind. A real surplus of scientists might emerge even if there were a stretchout—not a reduction—of these governmental programs.

At the same time we have not studied how efficiently these people are in fact being used. In one case it was estimated that the effectiveness of utilization of scientists in the operations of one of these large contractors was somewhere below 50 percent. Effectiveness here means any kind of meaningful involvement of a man's continuing energies in a project. There is real personal frustration built into the system of contracting when an individual does not understand how he fits into the larger scheme, when there is little feedback. This seems to be already an explosive situation and yet this has been the easiest decade. What of the decades to come?

THE MORAL ISSUE

Some of these issues have a moral implication. As people are educated, it is expected that they will become liberal, which implies that they will accept change. We have stated that it is management's responsibility to encourage such education.

We have described as possible solutions to the problems of economic growth that exist in the private sector, the development of the new types of investment opportunities in relation to urban development and in the fields of health and education, and welfare. This would require, however, a redirection of much of the traditional business attitudes toward activities in the public sector. The very fact that we can now question the size of the military budget is a symptom of this change in attitude. The fact that we have poured some $600 billion into an activity which has produced as its principal end product the ability to create instant destruction—with civilian and social benefits deriving only secondarily—is beginning to dawn on our consciousness as a problem of morality.

On a much smaller scale, the problems of management also concern issues which are ethical or moral—that is, concern with questions of ends as primary over means. In the past, management's job was primarily concerned with the means of accomplishing rather straight-forward ends. The size and number of organizations, the pattern of population, and the role of the government were such that the "invisible hand of the market" and "survival of the fittest" guaranteed a quick feedback between ends and means. Now we have the stability of "largeness" with at the same time a greatly enhanced means of accomplishing alternative ends. It seems that the problem is to change the management viewpoint so that it considers ends more explicitly. For instance, what are the purposes of organization? For what is so much human energy being spent? We need generalists in management because generalists ask this type of question, i.e., the moral questions. A new breed is emerging in management. This aspect dwarfs the problem of whether we should have scientists or non-scientists in management.

Another problem is that of the telescoping of technological time, the time between discovery and application. In the past much of the scientific effort developed as a result of curiosity. We are at the end point of such use of science. We must make a search for a new morality. It is not so much technological change as moral change that we are searching for.

OPEN QUESTIONS

1. The changing composition of the work force in the highly sophisticated and complex aerospace companies has resulted in steady progression toward a higher and higher ratio of scientific and engineering specialists vis-à-vis production and unskilled workers. In these companies the unskilled worker and even the production worker is falling into the same category as the agrarian worker in our nation's economy. While their role remains highly important, they represent a dwindling percentage of the work force. This is a direct reflection of the increased emphasis on encompassing the entire spectrum of scientific and engineering disciplines. It reflects the individual nature and limited production runs of even those items of hardware generally considered volume orders covered by the largest defense contracts.

2. Is it possible to deduce from the experience of the aerospace companies the generalization that the work force of those companies which primarily serve the civilian economy will experience a similar change? If so, what are the social and economic implications?

3. Does the growing emphasis on scientific-engineering types of personnel and the increased sophistication of the technologies employed mean that the manager as such will have a role of declining importance in the economic and industrial phases of our life? Does it mean that the manager of the future will necessarily be a scientist or an engineer?

4. It is apparent, at least in the aerospace industries, that the colossal research and development efforts of recent years,

which show no real signs of abating and which are largely government financed, are resulting in an acceleration of this technological revolution. Is it possible to effect a transference of these technologies, processes, and products generated primarily for defense purposes to the services of the civilian sector of the economy? If so, to what degree is this possible and what form is it likely to take? Assuming it takes place and assuming the rate of change accelerates similarly to the way it has in the aerospace industry, are the economic, educational, governmental, and social institutions sufficiently viable to adjust to it? If so, will the individual be able to adjust and does this apply to individuals at all levels of our society, or will there be marked differences in the capacity of individual groups to adjust?

With the changing composition of the work force in the aerospace industries, new phenomena are appearing. Two in particular are significant: 1) problems arising out of the nature of the defense business, in particular, the huge, long-lead time type of contracts; and 2) the limited span of productive years of the more talented, ingenious type of scientist-engineer.

The Dynamism of Science and Technology 4 *by William O. Baker*

Executive Vice-President, Bell Laboratories

This chapter will attempt to complement the previous discussion of some of the elements of science and technology.

Some things have to be believed to be seen and we will employ that technique during the discussion of some problems of science and technology. It is inevitable that some of the comments about the state of science and technology and about major currents in science and technology are going to be intimately connected to Washington influences.

Figure 1 is a pattern from coherent light and represents a dramatic example of the pace of modern science and the curious irony of its progress. We have thought we understood the wave properties of light since Newton's time, certainly from Maxwell's, yet this is a new kind of light which now seems worthy of the most searching study of the wave nature of light. It is quite possible that it has been lurking in some of the emissions from special configurations of neon signs! But recognition of this kind of light eluded us entirely throughout the whole history of physical science until the discoveries of Schawlow and Townes about three years ago.

Figures 2 and 3 present another approach to the impact of science on society, that of experimental medicine, with its consequences upon life expectancy, a scientific and technical incident of unrivaled import in the area of social problems.

THE DYNAMISM OF TECHNOLOGY 83

Related to population growth is the rate of growth of the number of scientists. Over the past few hundred years the employment of scientists has increased very steeply in Europe, even more steeply in the United States, in recent years at a

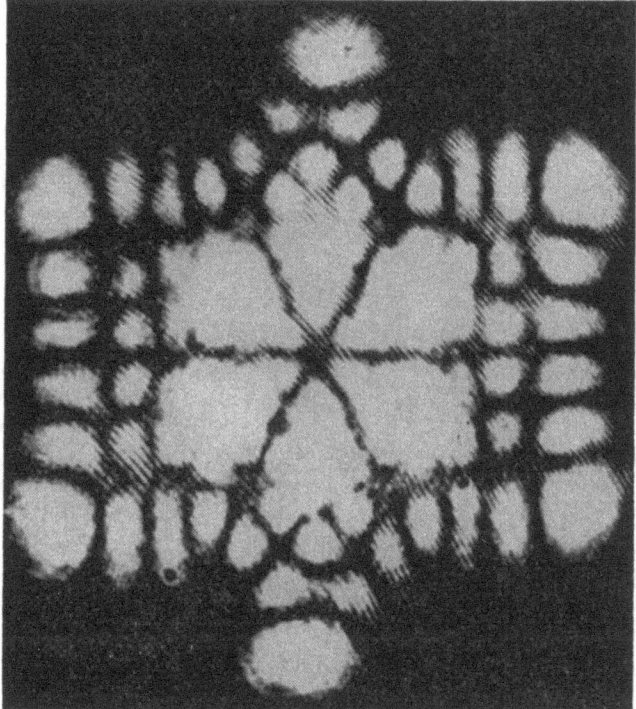

Figure 1. Pattern from Coherent Light.

nearly vertical rate (on a semi-log plot) in the USSR, and it appears that the rate in China, in a short while may exceed them all.

Can this sort of growth be maintained, or will it shift over to a plateau, as Professor Derek Price at Yale has suggested? Let us formulate the issue in the following questions: What are the current properties of science and technology? Are there going to be many more big discoveries and will these discoveries depend on many more people working in science and on spending much more money? Is the number of scientists in these countries going to continue much longer to increase at this rapid rate? Where are we on these various growth curves?

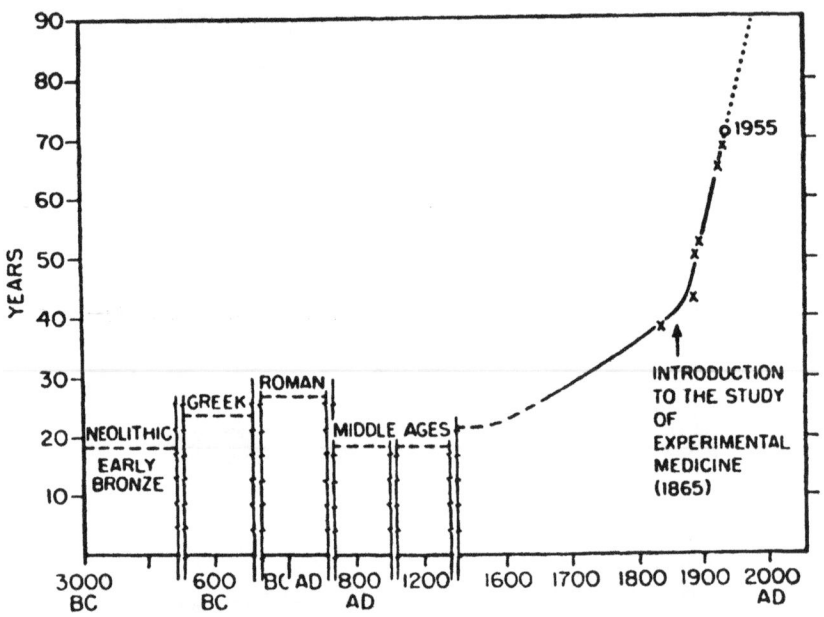

Based on M. L. Tainter. *Bell Telephone Laboratories, Inc.*

Figure 2. Life Expectancy at Birth (average of white male and female population).

Figure 4 shows that extrapolating trends is never entirely reliable. Consider the electrical manufacturing industry. In 1832, one individual, Faraday, almost established this industry; it grew, but not at a very fast rate until World War II. Then there was a hiatus, followed by a very rapid rate of growth which if it were to continue until 1990 would involve the whole working population of the world in that industry. Even according to its early growth rate, the number of persons who should be now working in this industry is far in excess of actual employment in it.

Related to the question of scientists and discovery is the narrowing interval between discovery and application mentioned in previous chapters. This is illustrated in Figure 5. There has been in the recent past an accelerating pace of discovery which involved an increasing number of people both in discovery and

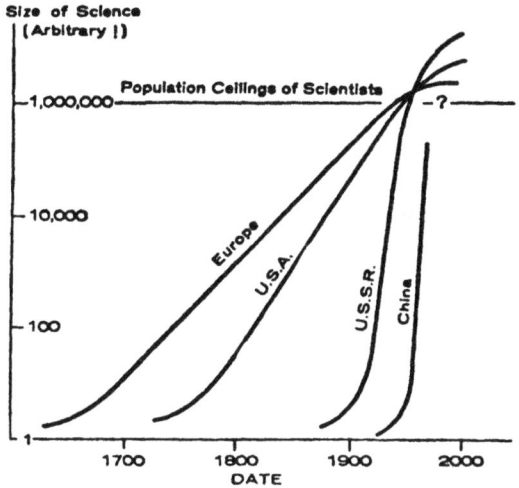

Figure 3. Rate of Growth of Number of Scientists.

in development work. There is a very rapidly narrowing interval between discovery and application. Of course, it takes a great many more people, almost ten times as many, to develop a process or product for use than it seems to require to probe a field and make a discovery. In the concepts of discovery and application, there is, as we have mentioned, a certain sizable amount of overlap because of loose definition.

Some of the recent discoveries illustrate these points. Almost eighteen months elapsed between the discovery of coherent light and putting it in a form suitable for application. To develop the super-conducting solenoid magnet, a quite remarkable and practical discovery, in a very practical engineering way, took a similar period of time. One of these innocent-looking magnets can be used to create a magnetic field thou-

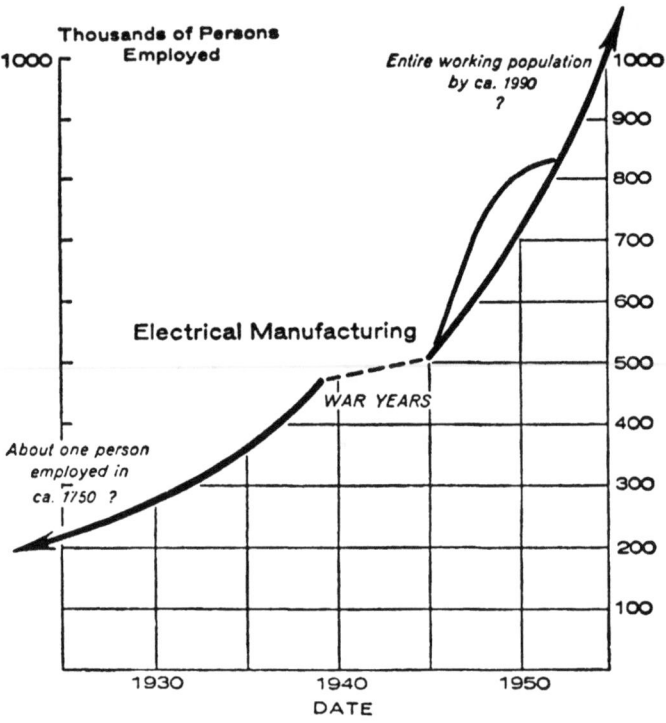

Figure 4. Rate of Growth in Electrical Manufacturing Employment.

sands of times higher than anything its size could previously create and at immensely reduced power consumption. If we eventually control nuclear fusion, or certain applications of magneto hydrodynamics, either of which could revolutionize the power industry, it will probably be through devices like these.

THE DYNAMISM OF TECHNOLOGY   87

Not only is the interval between discovery and application narrowing but in some ways the magnitude of the effects, the dimensions of the event, are becoming much larger. In this process there is a conversion of science and new discovery into engineering, and a feedback of the results of engineering into science through systems studies which guide the way that the conversion of science to engineering in turn must take place. A beautiful example of this interaction between scientific dis-

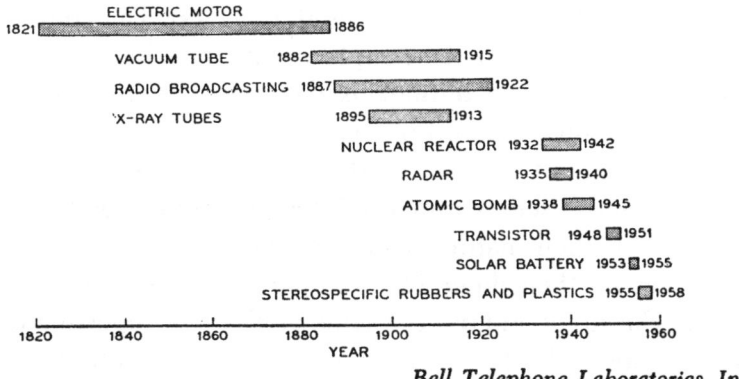

Figure 5. Narrowing Interval between Discovery and Application in Physical Sciences.

covery and engineering is provided by the nuclear energy program. In the late 1940s it was thought that nuclear energy would create a great deal of cheap power and that this would revolutionize industry. So far it has not had such an effect. In 1962, on a base in Antarctica, all the heat was turned off because every bit of the auxiliary heating capacity of the station had to be used to keep the nuclear reactor from freezing! The reactor itself was not workable and there was tremendous danger that the cooling system would freeze up and break apart. This is a late 1962 commentary on science and engineering in that field.

To use the kind of sophisticated science which we now have,

we must have systems research and systems development and systems engineering. In our society there is a curious impediment to this, produced by what we might call the antitrust syndrome, since the new systems often require complicated, expensive, and even antitrust types of behavior. Therefore, in spite of the fact that the major new scientific discoveries seem to require new systems for their effective application, there are numerous illustrations of the fact that we do not do very well in this respect. A question therefore arises: Must new systems be increasingly initiated by federal agencies? The communications satellite is a good example of this problem. Presumably the whole economic outlook is dependent upon how we approach and solve such problems.

Innovation, of course, is the primary mission of applied science, but it seems unnatural for governing agencies to innovate. By their nature such agencies attempt to regularize behavior. By their dicta, for example, we all drive on the same side of the street and use the same methods of reporting income taxes. Those who have had some association with the federal government know that it is not easy for government agencies to sponsor innovation. All the funds that are wasted, the cutoffs, the letdowns are understandable because government activity does not lend itself to the kind of risk-taking, the quick decisions that innovation through science and technology best thrives on.

However, government funds are deeply involved in research and development activities. Figure 6 illustrates the increase in such expenditures in the recent past. In fiscal 1963 the sum of such expenditures will crack the $15 billion mark and will be a little higher in fiscal 1964. The composition of this total is revealing. The life sciences have received a fair amount of these monies. In fiscal 1964 the National Institutes of Health alone will spend about three-quarters of a billion dollars. Other sciences are able to operate on a more thrifty budget. The department of government that spends the overwhelming proportion is Defense. And of course NASA spends a considerable amount on research and development.

The proportion of such research-and-development funds con-

National Science Foundation

Figure 6. Trends in Federal Funds for Research and Development.

tributed by business on one side and the government on the other shows great variation from one industry to another. Aircraft and missiles are of course almost entirely supported by federal funds. In the category called electrical equipment and communications, the electronics segment is federally supported while the communications component receives relatively little federal funds. The average for all industries is that 70 percent of all research-and-development funds are now contributed by the federal government. The field of professional and scientific instruments is beginning to show a more equal distribution of funds supplied by industry and the federal government. The proportion of such funds contributed by the federal government decreases as we move from wood products, to machinery, rubber products, motor vehicles, transportation equipment, chemical and allied products, textiles and apparel, mining, primary metals, stone, clay and glass, food products, and paper products.

One of the more serious conclusions that emerges in an analysis of expenditures on research and development is that those industries which are in some sense the tissue and fuel of society—the primary metals, stone, clay and glass, and food products industries that create much of the basic welfare of man—are simply not engaging in a large amount of research. This is illustrated by considering the number of scientists and technologists working in these industries. For instance in stone, clay, and glass industries, which includes the cement industry and is the backbone of all the construction in the nation, there are only about 4,000 technologists working in research and development and most of their work is not really scientific in character. Again, in the primary metals, there are only a little more than 5,000 technologists, while in ferrous metals—in a nation dependent on steel in every form—there are only 3,000 technical personnel. This is a trivial number in a nation of 188,000,000. The same observation might be made about the motor vehicle industry.

Turning to those industries in which there are large numbers of scientists and technologists, we find that out of the national total of above 320,000 such personnel, 91,000 are employed in aircraft and missile industries. About 50,000 are involved in military electronics and related subjects. There are, then, some 140,000 scientists and technologists in these two areas alone.

A recent government report identified textiles as one of those civilian areas which need technical help, but the industry only employs 1,200 technologists. This seems to be a pretty clear judgment by our society: if we maintain only that level of technical resource for the whole of a great industry, we do not want much more of it.

The dollar expenditures on research and development in the various industries confirm the conclusion suggested by the investment of manpower resources. Textiles, fabricated metal products, paper, stone, clay, and glass made very small expenditures indeed. The chemical industry of course spends substantial sums for research and development, primarily from company funds.

There is a question about the methods which would insure that we get some value from the research and development effort revealed by the above figures. If we doubled the number of people working in the ferrous metals, would we reinvigorate and drastically change this field? What can our expectations for science and technology be? Will they be somehow proportioned to the effort put in? Do we know how to create, or better control, the conditions which will foster big changes in science and technology?

An example to illustrate this problem is the work done by Harold Black which laid the basis for electrically controlled machines, or automation, a development of enormous consequence. This work was not done in a laboratory with flashing lights and elegant white-frocked assistants. It was done on an obsolescent and tired ferry boat. It was the result of a long, almost mystical, process of contemplation, experience, and stimulation. It was done under as nearly uncontrolled conditions as possible. Yet the results of this work are now very current. One of its most recent applications is in the work of biochemists who during the past year or two have found that there may be very strict analogies between Black's theory and concepts of quantitative analysis and the nutritional behavior of cells. This in turn underlies metabolism and the whole process of growth and of form.

Another illustration of a slightly different aspect, but again emphasizing the uncontrollable feature of the process of discovery of large new things is the work of Janski in radio astronomy, a discovery which is essential to space communica-

Bell Telephone Laboratories, Inc.

Figure 7. Early Janski Apparatus for Work in Radio Astronomy.

tions and ultimately to whatever economic activity there may be in space. This discovery is also related to many new and quite sophisticated kinds of microwave communications that may be quite important economically. Janski made his discovery several decades ago with very simple apparatus, much of which he constructed or had his assistants construct right on the scene (Figure 7). He went as far in identifying the sources of emanations from the stars as the physical theory of

THE DYNAMISM OF TECHNOLOGY

the time permitted. It remained for Purcell to discover the 21 centimeter wave length emanation from the hydrogen magnetic moment inversion spectrum, but much of these elements of physical science were not known at the time of Janski's discovery and it was not possible to make a full application of the discovery then. At the same time it has become increasingly expensive to pursue great simple ideas like this. It now requires

Figure 8. A Satellite Communication Antenna.

immense orders of effort, expense, and elaborateness to arrive at substantial new discoveries.

The best and simplest and cheapest radio telescope now costs between $30 and $40 million. The contrast between these devices and Janski's original instrument is tremendous. Figure 8 shows a simpler embodiment of Janski's idea, the first satellite communication antenna capable of billions of times more sensitivity than Janski's device, involving the most sophisticated sort of liquid helium operated microwave mazer, involving some very nice mechanics, and a very elegant computer system which points it and calculates orbits of the satellites which it follows. This is a very small model of the one that was subsequently used in Maine and in France. There seem to be no natural limits. We go further and further in accumulating and integrating the findings of science and technology of earlier periods and using them in the next step in science. The judgments that control how far we should go, given time, ingenuity, and intelligence, are not likely to be scientific ones.

We have by no means begun to saturate our study of the possibilities even in such deceptively simple things as the composition of matter. There has been great emphasis lately on solid state physics. In the late 1940s it was generally felt that no respectable physicist would spend his time on solids because they were hopelessly complicated, there were too many atoms together. It was held to be essential to get the atoms apart and look at them separately. There has of course been a tremendous change in outlook. All the matter of earth is now submitting to a great new invasion of understanding, already greatly advanced, and there is no visible limit of this work. For instance, the stability of matter, changes in its state, the ranges of its behavior, for example, with changes in the atmosphere. We now have practical conditions wherein we have to deal with matter at room temperatures and also at high temperature states. Space reentry is one example; modern machine operation is another. The lifetime of matter that we have to modify and control ranges from something like 300 billion years to 10 to the minus sixteen seconds. These are dimensions in which man has had almost no experience, but the problem has become a very practical one. It is not just some study in astrophysics. Our current science and technology, our current industry and economy, have to deal with these problems. The challenge is very great but so are the opportunities.

DISCUSSION

RESEARCH AND SYSTEMS

Research today requires a systems approach because the interactions among the technical elements of society are so strong. For example, if we were to try to introduce the transistor into the whole complex of radio, television, automatic controls, communications, without adjustments, without systems studies, it would simply perturb the situation and would not work out.

However, as we plug machines and people into a system, more and more pressure arises and therefore changes become more ramified. One problem relates to the system itself. For instance, the innovation of synthetic textiles required new spinning processes, new weaving processes, new darning heads, all of which entailed large capital costs. Does an outer limit to how far and how fast we can innovate arise because of the wide ramifications built into systems?

Second, as changes become built into so many different systems, little breaks occur all along the line. We are unaware of some of them which sometimes accumulate. At what point does the system requirement take hold? Where in the discovery-application cycle does it become important? It does not, after all, require a system to produce the ideas which result in innovation. But even here we do need the interaction which a system may best provide—the experimenter, developer, and the scientist-creator.

This implies that every society, as it develops a higher state of technology, needs more and more planning and direction. How then do we move away from government when the whole logic of the situation is to move toward more planning, more pertinent information about the ramifications of the system, and more and more of a central role for some planning mechanism which is equipped to provide a chart of the ramifications of the change? This proliferation emphasizes dramatically and graphically the need for an enhanced role of government, even though for other reasons we might like to back away from it.

Both public and private government are set in motion. Perhaps we need the development of some kind of power which is able to deal with systems while de-emphasizing public power which may have limitations because of its inherent rigidity.

On the one hand there is the planning function, the leadership function, apparently involved in the very technology of the situation, quite aside from the sociology of it. On the other hand, we would not like to see this all done by government. This may, however, be the only solution because of some of the other factors.

Perhaps, however, we do not need additional planning for systems. People can plan, people can carry out the integration that is required if they have the incentive, whether it is applied from the outside, from another system, or whether it arises within the system itself.

The planning can take any of a number of forms, top down, uncentralized. It does not mean the spectre of the omnipresent, omniscient mind. It means having a greater awareness of the ramifications of change and attempting to meet them. It does not mean governmental planning in the narrow sense. All sorts of new kinds of organizations are emerging to play this role. For instance, the university itself is one of the great agencies of planning and control.

Perhaps we do not even need more science and technology. There may be a point at which the attempt to keep systems in phase becomes unsuccessful. For example, we apparently have reached the point in military technology where we are trying to do developmental work and build missiles at the same time. This becomes fantastically expensive. Another example is the prospect of building a disarmament system; it has been pointed out that given the rate of the development of weapons, by the time a disarmament system is built, weapons would have changed so much that the system would be obsolete.

The above of course is purely within the scientific framework. But the problem also extends to people. How much change can you impose on people before the social structure breaks down? Can we afford a much faster rate of change? This consideration seems to impose some restraints which suggest that we should be looking for the places where we can safely innovate and not simply assume that we need more innovation in general.

We can of course halt innovation by not spending $5 billion in any particular area. The curve of research-and-development expenditures is not a natural one. It is a forced development and it can be slowed down. Another illustration of the possibility of slowing change is in the use of depreciation allowances. If we want an industry to expand, we can make depreciation allowances more favorable to investment. If we do not want expansion, we could penalize the industry through less liberal depreciation allowances. The same mechanism could be used both ways. Why, for instance, should we make it easier for the steel industry to invest $1 billion in order to reduce the break-even point another two points. There is no point in investing large amounts in steel if we have unused capacity.

OBSTACLES TO INNOVATION—THE ISSUE OF SIZE

We have stated that it is "unnatural" for government to sponsor innovation. This derives in part from the phenomenon of size. However, business is also hindered in integrating its innovations by size and by the "antitrust syndrome." For example, the size of the Bell System prevented the introduction of the transistor in its operations.

In addition, the depreciation policies and the regulation policies of government also block innovation. When government decides that it will be to society's benefit to subsidize or allow more liberal depreciation policies to hasten the junking of obsolescent plant and equipment, business will take advantage of this. Business itself has already done this without government aid in the solid state field.

There might be merit in forming small, new organizations perhaps under subsidy, which would try to find widespread use for new thrusts in science, rather than having the large organizations, either government and private, do it. The Commerce Department at present is trying to provide small organizations with more of a chance to use the new technologies; this is a sensible approach.

However, it may not be possible for the small organization to innovate because of the systems requirement. For example, Zerox was invented by a patent attorney at the Bell Laboratories, but the subsequent history of this innovation involved large-scale development facilities.

No small company could go very far, again for instance, in the creation of synthetic textiles. This requires a new spin process, new weaving mills, new sizing, new dyes, all of which takes enormous resources. But again some parts of the industry, felt making, mat making, and so forth, do not involve systems. There is no problem of perturbation in these sectors.

Some distinctions should be made. First, what is the scale of operations that is needed to support the initial discovery work of the scientist and technologist? Second, what is the impact of size on the exploitation of discovery and its application? Some very large companies have found it so difficult to exploit their own discoveries that they are moving to obtain leeway to do this both by alliances with outside firms and by internal divisions.

One solution might be to have more of the basic research conducted outside the large firms. If the results of the research are offered to the highest bidder, competition among firms would lead to the introduction of the innovation. This might provide some way of overcoming the inertia that might otherwise prevent a company from pushing research through the development and into the application stage. We make a distinction here between large companies setting up small groups and the small concern which is not allied with larger and powerful backing.

The small units are tied with tremendous tenacity to central technical resources. This is disturbing because we would hope to get flexibility through small units and also be able to tap the great diversity of interest, intelligence, and ingenuity of America. But the tendency does seem to be in the other direction, toward the large organization.

A real contradiction seems to emerge. Because of the systems requirement an organization must have a certain critical size in order to arrive at an innovation. But once that critical size is reached, the organization is so stability-seeking that it cannot introduce the innovation.

It appears, then, that size is not quite as important in the basic discovery, but size is vital in the development of the systems which will make it possible to apply these discoveries. At one time the obstacle may be the depreciation rates, an economic obstacle; at another time it may be a psychological obstacle, the incapacity of the large institution to accept this kind of change.

The Defense Department is a huge organization, traditionally bureaucratic, lethargic, slow. However, the Defense Department was reborn, simply because all these primal tendencies were abolished by modern weapons. The whole force of fear, the greatest emotion of an enormous society, pushed the Defense Department over its antipathy to change. The Defense Department had no choice.

It has always been difficult to sell an innovation to the Defense Department when they would have to absorb all the costs of change. It would take a tremendous amount of leverage and demonstration before they felt they could afford to move. We might generalize from this on *a priori* grounds: If it were possible to make an experimental area or an initial application area less vulnerable to potential loss, there would be much faster acceptance of innovation.

It is sometimes the technical element and sometimes the human element which sets the limits. It can be the technology or the human element in the organization that cannot reverse itself and take advantage of the possibilities offered by innovation.

One of the basic problems is that we fail to see the distinction between research and development—expenditures for each, the number of scientists engaged in each, and the question of application. Certain innovations require great research-and-development expenditures, a great number of scientists, but they can be applied quite simply. The size of the unit devoted to application may be totally different from the critical size of the unit necessary for the research and development.

New industries show the largest expenditures. One important factor which causes this result is a psychological factor. Management in new industries is conditioned to innovate, they are organized to innovate. But in older industries, management is hamstrung by precedent, years and years of traditional exploitation, point by point regulation which is bound into the codes of every political subdivision.

For example, the construction industry, which is so backward, is the quintessence of a systems approach. An individual cannot have shingles put on his roof without this having some influence on the fastenings, and the fastenings, in turn, on the joist. The position of the joist influences plumbing, which influences "x" through "y" and so forth. At the same time there is an age-old situation in which the specialist in putting on stones and bricks pays no attention to the wood fastenings, electrical conduits, plumbing, and so on. There is no systems study, no systems principle in the industry. As a result, if one tries to introduce an innovation into the technology of the industry, he will always perturb the rest of the system. No group is willing to look at the industry as a whole, and consequently nothing can happen in the industry.

A study of the construction industry was started for a report on civilian technology. Even the preliminary assessment of the industry was never finished because the multitude of inhibitions on the most primitive systems study was overwhelming. For example, no building operator, no building company can put up a structure independently of craft union influences. There is no piece of land on which a structure can be erected without the craft union's directing how it should be done. The construction industry cannot even undertake an experiment.

Do the craft unions, then, represent a dead stop to building innovation? Not really. First, economic pressures from abroad have not yet had their real impact in America. Second, both management and the craft union have to learn the necessity of some give and take on the issue. But there will be a lot of strain and rigidities in the system, and cracks will appear, before people abandon their entrenched positions. It will take an enormous amount of pressure.

One way to get a corrective mechanism in the building industry would be to reach a crisis. If we experienced either a fantastic population explosion or a fire which would destroy a quarter of our houses, we would probably alter the situation.

Banks and insurance companies, characteristically thought to be some of the most bureaucratic of organizations, have had a crisis and have been forced to innovate.

After World War II the Germans revised their construction industry. They drastically improved, although they did not completely revolutionize, their techniques. When NATO decided to set up a science advisory group, the Germans sent an architect-city planner as their representative because he was the most revolutionary of the possible candidates. And he was used to dealing with systems.

The Germans lead in the study of pre-stressed concrete because they were forced to innovate. They also eliminated craft unions in the industry. There is a common wage scale and one does not need a card; anyone can overlap into another individual's work. This was done because of the kind of labor which was available.

However, to relate all this directly to a crisis situation is a little too simple. The Russians have had a similar crisis in the building industry and they have not been able to do much about it.

Today societies which are backward or underdeveloped in certain areas are carrying on research and development in industries in which we have ceased to have any large interest. We are no longer training mining engineers or engineers to lay railroad tracks or build locomotives, but the Russians are. The whole level of the technology and the society is involved here. There is much more of a social relationship involved than is presented by the simple technological facts.

AGRICULTURE AND INNOVATION—SIZE AND SYSTEMS

Agriculture provides an extreme case of contrast. American agriculture has been very adaptable in the face of changes in technology unparalleled in any other sector of the economy.

How is this consistent with what has been said about systems and size? Who could be more resistant to change than an old farmer who is used to farming a certain way? Yet here, without a question of a doubt, we have the greatest adaptability. Here is a case where there has had to be a tremendously complicated system of adjustments. People have had to change food habits, to change their ways of growing food, of dealing with the market.

This adaptive process was made possible through an immense apparatus of government agents going into each area and working with the farmers in an extraordinary way. It was possible to demonstrate clearly the differential between the old and new; people wanted to escape from the old, particularly hard unpleasant labor, when they saw the possibilities of the new way. The farmers' situation was compounded by the loss of their manpower and their womanpower as their sons and daughters migrated to the cities. The farmer really had no alternative.

What then are the orders of pressure which make people accept the new? It does not seem to be a question of size as such. It is more the question of demonstrating clearly and concretely the cost-benefit equation. We are even going to see adjustments in the race issue in the South because the cost-benefit system has changed enough to cause the blow-up of a tremendously complicated system in a remarkably short time.

Behind the agricultural revolution has been a tremendous network of systems research, primarily government research in the agricultural experimental stations. The most dramatic embodiments of it have occurred in the past fifty years, although land-grant colleges and extension services have been in existence even longer. The fantastic agricultural developments of the 1950s, since the development of the new chemical agents, have been in turn subsidized by agricultural surplus and subsidy policies, which constituted a set of experiments and development, particularly development, which dwarfs the space program both in cost and scope. Farmers did not act as individuals. They were enormous groups who were led by extension agents, county agents, field agents. The systems experiments within this particular area, in machinery, in chemical crop agents, fertilizers, herbicide control and so forth were also of enormous significance. The only way this agricultural revolution could have been accomplished was through a government sponsored system. Many individuals were involved and it covered a very big geographical area. The job could only be done through government agencies.

In addition, it was the first, the wisest, most enterprising recognition of the unique role of the university in modern society; that the universities can create a resource. This was not done in communications, steel making, automobiles, even nuclear energy, but it was done in agriculture.

Without minimizing the role of government in agriculture, it should be noted that we had a sizable agricultural revolution before the Department of Agriculture ever started, a revolution carried out in part by farm implement companies selling equipment to farmers. There was a tremendous built-in momentum in the market place of differential opportunity for the farmer. To the farmer the purchase of farm machinery made sense. It is true that government agents added a new dimension to these changes but large-scale technological change or social change does not occur primarily or solely through a big systems approach.

On the other hand, farm machinery was not very important until the development of lubricants of the oil industry had started. Farm machinery, on a large scale, had to wait until ferrous alloys had been developed with rust-resistant qualities. It was not possible to exploit a piece of agricultural equipment until all the technical variables had also been developed. This was done on a broad front. The agricultural machinery people did field tests; they worked hand in hand with metallurgists; agriculture thus became a systems development.

THE DEVELOPMENT AND ALLOCATION OF SCIENTIFIC PERSONNEL

With regard to the overall shortage of potential scientists, we know that there are a lot of untapped resources. The National Science Foundation tried, by new approaches to curriculum, by countless summer institutes and so forth, to locate the untapped potential in the high school. We have to say that we have not found it. That does not mean that it is not there. It does not even mean that our hope for approximately doubling the number of highly trained, qualified scientists and engineers is permanently doomed. We may have overlooked some factors in the area of motivation. To utilize motivation would mean a kind of bribery—that is, to get young people to go into science and engineering who otherwise would not. When the results of most of the programs are added up, there is not much evidence to tell us of a large untapped source.

The Institute of Personnel Assessment in California, which is concentrating its activity on studying creativity in individuals, has done very interesting work. The conclusion that seems to be emerging is that detailed educational assessments at the secondary school and college levels are not correlated with the "creativity" or objective proficiency of technical personnel in their ultimate performance as adults. There had already been a signal on this in studies made by General Electric.

If we are soon going to be pushing against the ceiling of scientific manpower, what kind of criteria do we bring to bear in the selection among alternative ends for the investment of scientific and engineering skills and talents? This seems to be one of the increasingly difficult problems in our society. Congress apparently will spend any amount of money for space. A number of large corporations have sufficient scientific manpower resources to insure their making sizable investments in their technology. But the question really is: How does a complicated society like ours get a chance to express preferences? And where does it get guidance in this area, since we are a democracy, if not from the scientist?

We do not have a good mechanism for deciding among alternatives. There are efforts being made in Washington, clustering around the President's Science Advisory Council, his special assistant in science and technology, the Federal Council for Science and Technology, and related groups. These groups have tried, and increasingly during the past two or three years, to determine what ought to come first in terms of national welfare and national desires. Three or four years ago a study of constitutional amendments which would permit the election of science members to Congress was proposed. This would at least allow a debate on these questions and the bringing of these issues to the electorate. A second possibility might be a special electoral system. A third possibility is the Russian example—to create a National Academy of Science to pass on the priorities.

The Vice-President of the United States, who is responsible for the space program, invited scientists to come to plead with Congress for the passage of the space budget. But for a group of insiders to tell a group of outsiders whether they should continue certain courses of action, without responsible discussion from other points of view, seems to be a circular approach.

By and large, the leaders of the scientific community have not confronted these problems. They have been so concerned with being useful to the United States and with "running" with their own ideas and programs that they are not doing the minimum job of making clear for the rest of us what they already know to be some of these difficult problems. Although the

American public has a way of catching up, the scientific community should play a larger part in the education of their fellow citizens.

This very factor has caused an enormous amount of distress, discomfort, and debate. There will be a serious effort in the near future to start laying out some of these priorities in forms which can elicit public understanding and debate. But this effort will have to face the fact that the first item in the budget which Congressmen usually slash is the section dealing with public understanding of science.

There is a great discrepancy between what we might call the civilian, or use, sector of the economy, and certain other sectors of the economy. In the civilian sector, it is difficult to express demands for innovation and for the expenditure of funds for research and development. In many instances, we do seem to know what we want and the demands are genuine. For instance, there have been sufficient large-scale developments in metals and other materials to have spilled over into the housing industry and surely speculative builders are as entrepreneurial as farmers in wishing to cut costs and make profits.

The problem is that builders do not have the systems development that we find extensively through agriculture. They may know what they want but they do not know how to get it whereas the farmer learned through his county agents, through the equipment and tractor salesmen, the technical services, of, for example, McCormick and Deering.

Productivity and Economic Growth

5   by Solomon Fabricant

Director of Research, National Bureau of Economic Research

This chapter will discuss the areas associated with productivity, technological change, and automation. We will first consider some accepted ideas relevant to these areas—the idea, for example, that technological change is occurring or has been occurring in this area; the idea that the consequences of these changes have been so great as to require a considerably different set of institutions, even a revolution in the management of economic affairs.

We will begin by presenting some of the many misconceptions that have arisen in recent discussions.

The first misconception is that productivity is a clear, simple concept which requires no qualification, added adjectives, or care in definition. This is wrong. Productivity refers to a family of concepts. Although the members of the family have resemblances, productivity or productivity change usually refers to some kind of a ratio of physical output to physical input, usually in an index number form. This is about all the members of this family have in common.

One productivity index which has been used for many years is an index of change in output per man or output per man hour to labor input. Another index which has been developed more recently, but is beginning to be more widely used and recognized, is an index of output in relation to labor of different qualities which are weighted according to their relative values. An index of output per man hour in which man hours are unweighted behaves of course quite differently, for many purposes vastly differently, from an index of output per weighted man hour.

Still more recently, statisticians have begun to calculate indexes of output per unit of total input, by which they mean output per unit of labor and tangible capital. When capital is taken into account and counted as an input, there is still a different concept, and quite a different index of change in productivity.

So far we have discussed only indexes of national productivity. For individual industries, and we are frequently concerned with indexes of productivity for individual industries, there are still other variations of this concept, other members of the family of concepts. For example, the treatment of material input for the nation as a whole can be ignored, but the treatment of national input for an individual industry must be carefully considered. It can be ignored in most of the indexes of productivity for individual industries. This gives one result. But if material input is subtracted from gross output to get a corrected measure of output, to give a net output per unit of input—to be a little technical, sometimes we say a net value added, per unit of input—a quite different index results. If an index of productivity were calculated in which materials were not subtracted from the numerator, output, but added to the denominator, labor and capital, to give gross output per unit of labor, capital and material input, still another index would be derived.

The Department of Labor, governmental agencies, and the National Bureau of Economic Research have been making calculations of this kind, and a variety of such indexes result.

In talking about productivity, and in using measurements of productivity, it is essential to be quite explicit since these different indexes have different uses. They are not substitutes for one another, but they are complementary in some degree.

A second misconception is that an index of output per man hour, such as obtains by taking the real national product and dividing it by man hours, is an index of efficiency. A better index of efficiency would be the real national product divided by total input, including capital and labor input.

The reason is that output per man hour may rise, has indeed risen over the years, because of increases in capital. Capital is a substitute for labor. To say that an increase in output per man hour has occurred may mean an increase in efficiency. However, it may mean no more than substitution of one kind of input, capital, for another kind, labor. It may mean, and usually has meant, a combination of both.

Therefore, output per man hour generally overstates the increase in efficiency, as we think of it in an economic or even in an engineering sense. Output per unit of total input, taking account of capital input as well as labor input, is a superior measure of productivity, in the sense of efficiency.

This is at the fringe of current practice in this area. There have been some efforts at broadening the concept of input to include not only tangible capital, plant, and equipment, but also to include education, for example. An index wherein a unit of output is measured against a unit of weighted man hours is a step toward taking account of the capital that is invested in human education.

To press this further, one might attempt to count as input, and include in the denominator of the productivity ratio, investment in research and development. This kind of calculation has not yet been made, but it is under discussion.

Another erroneous notion is that the long term rate of growth in output per man hour in a country such as the United States has been about 3 or 4 percent per annum. This does not mesh with the facts at all. The best estimate is that over the last seventy-five or hundred years the long term average rate of growth in output per man hour in the United States—and this is probably the highest long term rate that any country has had for any comparable period—has been of the order of 2.3 percent per annum.

One reason for this misconception is that often "long term" is held to mean but a few years. However, the period of time must be about seventy-five to a hundred years since productivity change does not occur smoothly. The process of technological change and all the other factors that are represented by productivity advance are not part of an automatic, smooth process. The process is a fluctuating one for the economy as a whole and even more so for individual industries.

With regard to the United States—but this would be true of other countries—the figures for as far back as the record goes, to the Civil War, indicate rather long periods (ten, fifteen, twenty years) in which the rate of growth of productivity has been high, considerably in excess of the 2.3 percent referred to. There have been other periods of ten, fifteen, or twenty years in which it has been low, less than the 2.3 percent. There is no constancy.

In addition, of course, there are cyclical fluctuations, short term fluctuations associated with the business cycle, for example. Thor Hultgren's recent study done at the National Bureau of Economic Research, *Changes in Labor Cost During Cycles in Production and Business*, throws some interesting light on the character of the cyclical fluctuations that occurs in output per man hour. There is a systematic pattern of change in output per man hour, from the trough ending one recession to the peak, and then to the next trough.

In addition to short cyclical fluctuations and the longer

waves, there are irregular movements, even in the national productivity index. One reason is that there are very real random factors affecting productivity change. These factors include the weather, strikes, and differences in the rates of introduction of new technologies.

The irregularities also reflect statistical error. The productivity index is an index based on a ratio. It is output in relation to input, or output in relation to man hours. The index of output is not a perfect index. The real GNP figures put out by the Department of Commerce are subject to error which sometimes are not negligible. We cannot be sure what the change from one year to another has been in the real GNP, even after they have been revised.

The man hour figures are not perfect either. Indeed, the Department of Labor, which puts out the official indexes of national productivity, has two indexes for the private sector, one based on man hours derived from the monthly survey of the population, the other obtained from established reports sent directly to the Department of Labor. The difference between these two series of man hours is sufficiently great to make it desirable for the Department of Labor to present two series. The Economic Report to the President shows both series in the appendix.

Another misconception is that productivity has been subject to acceleration, that we are living in an era in which productivity rise is appreciably, if not vastly, greater than it has been. Actually, the figures for the United States, which are as good as there are, do suggest that since World War I, the rate of increase in output per man hour, or several of the other measurements of productivity, has been somewhat more rapid than the average rate of growth in the period prior to World War I.

But to jump from that statement to the statement that there has been acceleration—that is, that from one period to another there has been progressive increase in the rate of growth of productivity, particularly in labor negotiations—is wrong. A few years ago, the trade unions argued that output per man hour in the United States was accelerating, and held therefore that in labor negotiations the relevant figure that ought to control the improvement factor was not the past rate but the prospective rate, and in accelerating productivity, the prospective rate is naturally higher than the past rate.

The Department of Labor, in its publications on productivity, actually fitted appropriate trend lines to the statistics, which had an acceleration factor in them, and they seemed to fit reasonably well. However, the statistics which the Department of Labor used covered the period from about 1909 to 1955; if these were extended to 1900 or 1890 or earlier, a trend line would not indicate any acceleration. Perhaps more crucial is the fact that the statistics do not indicate acceleration.

A related conception, that there has been a revolution in technology, is also wrong. A more correct formulation is that there have been both a revolution and no revolution. There is always a question of degree. But to say simply that there has been a revolution in technology is not justified. Perhaps there will be a revolution, but we have not yet seen one, according to the records on productivity and all the other relevant statistical records of the economy as a whole.

Another misconception is that of equating productivity change with technological change. Productivity change of course reflects technological change. Technological change is an important factor in making for an increase in productivity. But it is by no means the only factor. Reference has been made to recent experiments to calculate investment in education as a form of input, an aspect of capital investment. Clearly an increase in educational investment—more properly an increase in the stock of educational capital—will have some effect on the rate of increase of productivity. But this is not synonymous with technological change. Surely research and development, of which so much is being made these days, is a factor affecting productivity, but it is not synonymous with technology. Economists talk about the economies of scale, which refers to the fact that the United States, because it is a big country, without tariff barriers at the Hudson River, or at the Mississippi, can be more efficient than small countries which have obstacles at their borders. To the extent that there are economies of scale, one of the reasons the productivity in the United States is large today, larger today than it was years ago, is because we are bigger economically.

Another explanation of why technology and productivity are not synonymous, why increases in productivity do not necessarily depend on technology, is the fact that productivity depends also on what might be called the character of consumption. Education can be viewed both as a kind of consumption and as a type of investment. We might try to separate the two, but there are types of consumption in which it is very difficult to think of an investment factor as playing any part in the calculations of people. For example, whether people like to play games is a factor, in some sense, which makes for a difference in productivity. This is a relevant factor, which surely cannot be subsumed under the category of technology.

The economic policy of a country surely affects the output per man hour or output per unit of labor and capital. Economic policy can be better or worse, but it is not technology. Indeed, if economic policy does not affect productivity at all or appreciably, economists have been wasting their time for the last few hundred years, because precisely the things they worry about are such things as the tariff, restrictive work practices, monopoly, and competition, all of which are not technology.

Then there are what economists call exogenous or autonomous factors, which are not technology. Yet they affect the movement of productivity. Professor Dolmar wrote a review not so long ago in which he took issue with the statement that output per unit of total input was a reasonably good measure of efficiency, or even an approximation of efficiency. He stated that if one took account of all the inputs, factors that have been ignored, like education, investment in research, and so on, except perhaps economies of scale, the productivity index would remain constant. The residual would vanish or effectively vanish. This may not be so but it is a moot question. It is partly a matter of definition. It is important to know what it is that creates growth, and this certainly includes the kind of things people consume and the kinds of economic institutions they have.

In order to measure the rate of technological change for the economy as a whole, probably the best measure, although not a good measure—is the productivity index. Of course there have been other measures. One of the measures of technological change used by German historians was based on a count of great inventions. This shows, for instance, the acceleration which could account for the industrial revolution. However, there is no explicit definition of a great invention. But all the bright ideas of Leonardo da Vinci were counted, and counted at the time Leonardo wrote them down. Think of the implications of that particular measure!

To return to the productivity index: if we use it as a measure of technological change we will find no acceleration, no revolutionary change. The rate of increase of productivity in the United States today is of the same order of magnitude as the long term rate of growth; it is approximately 2.5 percent. It fluctuates, but it fluctuates around 2.5 percent. In some countries of Western Europe and in Japan and Soviet Russia, there have been somewhat higher rates of growth recently. There is some question as to just what these high rates mean, whether they are temporary results of the war, or of the postwar reconstruction or whether they are permanent. This remains to be seen.

But certainly as far as the United States is concerned, there is no clear evidence that there has been a revolution caused by technology for the country as a whole. In particular industries, in particular products, of course there have been revolutions. But even with regard to particular industries, there has been some tendency to exaggerate.

When speculation began in 1946 about the promise of atomic energy, it was said that in ten or fifteen years Consolidated Edison would have converted over to atomic energy, or would be bankrupt. But the recently published study by Resources of the Future, which presents projections on the production and consumption of various materials, fuels and sources of energy, and in fact makes extrapolations to the year 2000, indicates that it is expected that among the sources of energy in the United States in the year 2000 atomic energy will account for only a relatively small fraction.

Another common misconception is that productivity increases, even technological changes, have been largely concentrated in particular sectors of the economy. This is erroneous; technological change, or productivity change, to put it more generally, is a widely diffused process. Of course, in a specific industry at some time there may be very rapid increases in productivity. But to jump from that to the assumption that in the rest of the economy there is no increase in productivity is a mistake. In fact, one of the most interesting results of the statistical studies that have been made is that increases in productivity occur in every industry for which there are adequate records for periods long enough to eliminate the little wrinkles which occur in any of these series. It is hard to think of an example in which that has not been the case. Even government surely has increased its productivity. More needs to be done in this area, and a large scale study of productivity in the service industries will be made at the National Bureau over the next few years.

The few indexes that we have which seem to suggest no increase in productivity usually ignore quality change. For example, if one were to try to measure the output of the locomotive industry when it used to produce steam locomotives, by the number of locomotives produced by man hours, there would be no increase in productivity. But it is a very simple and obvious fact that the locomotives that were being produced toward the end of the 1920s were vastly different from the locomotives being produced around 1900. If one took account of that quality change, by various simple or not so simple methods, a clear increase in productivity would be indicated.

It seems clear that productivity increases are being manifested in every sector of the economy. This, of course, has great implications for policy. It means that policies that stimulate productivity have a very wide front on which to operate.

Another misconception concerns the factors responsible for the increases in productivity. In this case some misunderstandings arise from the use of words. Output per man hour, which is the oldest statistical measure of productivity, is usually called an index of labor productivity. Many interpret the term "increase in productivity" to mean an increase in output per man hour which is the result of some special effort of labor. It is argued that this of course has implications for wages and so on. But this again is wrong. Output per man hour may increase because of capital substitution for labor. It may increase because of management's improvements in the operation of the plant. It may increase because the labor is better educated, better trained, healthier. It certainly is not entirely or necessarily due to labor. Indeed, it might not be due even primarily to labor. Many factors are involved in accounting for increases of output per man hour.

The same sort of misconception arises from the figures on growth and productivity in individual industries. It is sometimes said that the productivity of, say, the automobile industry in the United States, the number of automobiles produced per man hour, has shown a great increase and that this reflects the efficiency of that industry. Of course, in part this is true, but in very large part it is not true. One reason, for example, that there is greater productivity in the automobile industry today compared with years ago is that the steel industry has improved its steel. A press can now be used for making an automobile body, or the several parts of a body, and this eliminates entirely all the labor that used to be required in the 1920s and earlier, to bolt and weld together a lot of little pieces of steel. Earlier it was not possible to use the kind of presses we use today because the steel available could not stand the necessary pressure.

Was the steel industry responsible for the increase in productivity in the automobile industry? Or was Columbia University, which contributed research and development to the metallurgical operations in the steel industry? Each was responsible.

It is difficult to identify responsibility. One can say that the increase in productivity in the United Kingdom or in the United States does not necessarily reflect the efforts of the people resident in that country. Some of the ideas for the automobile were imported from France. A lot of the ideas and equipment are imported and exported.

A related misconception arises in connection with labor, wages, and productivity. It has been argued, for example, that if labor gets an increase in real wages which runs parallel to national productivity, labor will be gaining all the advantage due to the increase in productivity. But again this is wrong. The simple arithmetic of it is that if labor is getting 75 percent of a pie and the pie doubles in size, and labor's share rises proportionately, it will still get 75 percent of the pie. This is not all of the increase. The other 25 percent is also absolutely doubled. This kind of argument has been used against the guidepost proposal made by President Kennedy that if wages were geared to national productivity, labor would get all the increase.

A related misconception is that wages in an industry ought to be geared to productivity of the industry. This too is fallacious. It is true that in industry in general productivity has been rising, but at diverse rates—in some industries, rapidly, in some, slowly. In those industries in which productivity has been rising slowly, if wages were to be geared to an industry's productivity, the wage scale in that industry after a short period of time would be completely out of line with wages in similar occupations in other industries. Wages in the electric light and power industry would be about 150 times larger than wages for the same kind of labor in, say, the lumber industry in the United States. They have vastly different rates of growth of productivity. The appropriate index, if one wants to use a general index as a guidepost to wages, is national productivity.

Another misconception is that increases in productivity or the introduction of automation—and automation and technological change have been used as if they were synonyms—are creating a vastly difficult problem. But the problem is exaggerated. This does not mean that it does not exist. Of course it exists. But, first of all, those industries in which productivity has risen most rapidly are the industries in which employment has risen most rapidly. There are of course exceptions but there are any number of examples: automobiles, since 1890 or 1900; television since the postwar period; rayon since the 1920s. Those industries in which increases in productivity have lagged most are usually the industries in which increases in employment have lagged most or in which employment has actually fallen. There is not such a simple connection between the rate of increase in productivity and the rate of unemployment as might appear to be the case.

One reason that a rapid increase in productivity might mean more jobs is that a rapid increase in productivity means that costs can be reduced and that under the pressure of competition prices will fall. We are still a competitive society, and when prices fall there is an increase in the demand.

Further, an increase in productivity, not only in a particular industry but in the economy at large, affects real income. Increases in productivity mean increases in real income. Increases in real income mean increases in demand. Consequently, even those industries in which labor would be displaced, if output were to remain constant, will not displace labor because output grows in response to an increase in real income and in demand. Most industries follow this pattern.

In those industries in which changes in productivity are of such a character, and demand is of such a character as to lead to a decline in employment, very often there is no problem because people are not pushed out of the industry. Rather they are attracted out of the industry by better opportunities elsewhere, in other crafts, other industries.

The problem of automation really concerns only a part of the employment picture, that part usually consisting of older people, the unskilled, and so on. In particular industries in which a revolutionary change due to automation is actually occurring, in which jobs are being lost, there is a problem of retraining and resettlement. But in general we tend to exaggerate the effects of automation and technological change and productivity change.

One reason we tend to exaggerate it is that we forget that

people want goods and services, and there is in fact no visible limit yet on demand. It is an old question—what are we going to do with all the automobiles? What are we going to do with all the products? The answer is that people want those automobiles and those products.

In 1850 it would have been hard to understand why, a hundred years later, people would not reduce their working time to virtually zero in light of the increase that had occurred in productivity. How much more beef and pork and corn and apples could they eat? How many more muslin shirts could they wear? How many more oxen could they use? What happened, of course, is that there are new products, better products. Moreover, people want to work. In the United States, despite the considerable increases in productivity that have occurred in recent years, hours of work in manufacturing industries have not fallen for over twenty years. We have, not fewer women in the labor force, but more and more women in the labor force. We have moonlighting. People want goods and services, and there seems to be no end to their wants. Employment will not vanish, though hours of labor may be reduced still further over the years.

There has been much erroneous thinking to the effect that the world is so utterly different, technological change is proceeding so fast, automation has taken such a hold, that we need an entirely new management of our economic system. It is said that government's role, particularly, needs to be quite different from what it has been. This is a very broad statement. There is a very important role for government, but that role is to help increase productivity, because the American people want to produce more goods and services and they want to provide more aid to other countries. Government has a very important role in increasing and maintaining competition, both in industry and among labor organizations. Government has a very important role to play in increasing incentives for labor, management, investors and savers, by reducing taxes and reforming the tax system. There is a great deal of time and energy spent on attempts to reduce taxes, time and energy that would be better spent trying to reduce costs. Government needs to do more about research and development, more about education. Whether it needs to spend more money, or whether it needs to direct more efficiently the money that is being spent, is a very serious question.

The government needs to spend more on transportation. Whether these should be an interstate highway system, of the kind that is flourishing in Vermont, or whether it ought to be concentrated on suburban and intraurban networks is another question. Another example of the need for increased government role are the building codes. They are obsolescent and need revision.

In Europe productivity agencies have been set up. Britain has a productivity agency, France has a productivity agency, even India has a productivity agency. But in the United States, we have no productivity agency. This is an example of a possible expanding role for government.

DISCUSSION

TECHNOLOGICAL CHANGE AND PRODUCTIVITY

We have learned that productivity indexes are an insufficient measure of the relation between technological change and social change. Economic averages, on the whole, over long periods of time, are equally useless in relating technological change to social change. The interstices and conflicts between job status, job loss, job gain, differential mobility rates for skilled and unskilled people, the increasing gap between skilled labor needs and the available supply are the very meat of the relation between technological and social change. Productivity indexes, therefore, cannot measure the relationship between technology and social change.

We have noted that technological change is just one element in productivity; similarly, productivity is but one element in technological change. As a matter of fact, many of the nontechnological elements in productivity—namely, the educational level, the nature of consumer demand, have some relationship to technology. For example, the nature of consumer demand might depend very much on what technology makes available.

It has been stated that there has been no revolution in technology, but that there have been revolutions in individual industries. However, if all the industries sustaining revolutions are added up, eventually there will be an industrial revolution which affects all of society. Historically, the industrial revolution in Britain in the eighteenth and nineteenth centuries did not occur all at once in every industry. It started out in textiles and metallurgy, then it spread to other industries. Our immediate historical evidence indicates that there have been changes in basic materials and energy sources and tools and machines and in the relationship of the worker to his job. Eventually this will add up to a new industrial revolution.

Perhaps it is production rather than productivity which brings about change. For example, the automobile assembly line did not constitute a revolution in American social mores when Henry Ford first ran the Model T through a production line. But at some point we had enough automobiles on the road to change our society pretty thoroughly, to change the whole outlook of our society, to make it into an urban society.

The same thing could be said about television. The productivity in television sets has not increased greatly, but at some point we reached a critical mass in television sets which changed the entertainment pattern of the nation. Perhaps then it is production rather than productivity which is responsible for some of these social changes.

The question is whether there has been sufficient acceleration to warrant the term "revolution." The steam engine of James Watt was revolutionary. The power loom was revolutionary, the railroad was revolutionary, atomic energy is revolutionary, television is revolutionary—we are going to continue to have revolutionary changes. But the problem continues. Its nature may change, and we have to revise our notions as to what is proper for government to do. It is not clear whether we have passed a critical point. There is a tendency to exaggerate the importance of the calculators and computers. Automation is not going to happen overnight. There are costs and profits to be weighed. In the long run we will have much more automation than we have now, and the people will be working at something else. But there will be problems of adjustment.

Do the elements of cost and profit determine that technological change will not happen overnight? We assume that as a matter of national purpose, we must rush full speed ahead, we must become as efficient as we can as rapidly as we can. We tell the workers who are affected, in any industry, that this is a patriotic obligation, that we cannot hold back progress. Trade unions have accepted this, but they ask for certain protections. We say that management has the right to determine, in terms of really narrow considerations,—considerations dictated by the market and determined by the rate of profit—when they will automate and to what extent and how fast. However, if it is truly an overriding national purpose to become efficient, can we continue to rely on the primary influence of the market or of private decision-making? In moving toward the full utilization of our resources must we not accept a new internal social standard, a much greater degree of centralized economic planning?

Among the people who are preoccupied with technological change and social change, there is a growing awareness that an endless capacity to consume more and more widgets and gadgets is not a sufficient justification for an economic and social system. The growing productivity, the technology, or the change in the range of gadgets that are available through this technology, the need to consume—all these are fundamentally unsatisfactory as a basis for a society to the people whose standards are now such that they are concerned with these problems.

THE NATURE OF THE DEMAND FOR GOODS AND SERVICES

The demand for goods at the average rate of increase of productivity in the past will probably not disappear for at least a century. However, although there are tens of millions of people in the United States who have far from a surfeit of goods, the dynamics of our economy require breakthroughs—new goods that people with disposable income may be interested in purchasing. There could still be a paradox; large numbers of people not employed because there are no new kinds of goods or services being produced and sold to people with disposable income.

Are we able to increase our rate of productivity growth, given the long-run picture of stability which seems to reside in the culture? In all likelihood. We can give more scope to the energies of private enterprise. For example, a tax reform and tax cut might enhance the rate of growth of productivity as well as the rate of growth of total output. We have mentioned productivity councils and agencies; support for research and development; improvements in education which could be generated. But it is not simply a question of spending more government money. It is a question also, perhaps even in larger part, of improving the environment within which the initiative and instinct of leadership that make for change, discovery, invention, addition to knowledge can operate well.

In addition there is a question about the nature of the quality of goods and the types of goods that an advancing economy wants. The three G's—General Motors, General Foods, General Electric—are simply not in a position to provide a wide host of the goods and services that the consumer wants to buy and has the money to buy. This has nothing to do with the rates of change as such; it is more of a qualitative dimension in the nature of the latent demand at this stage of society's progress. In metropolitan living today, the consumers' wants differ from what the private economy is prepared to supply.

In addition, there are problems of institutional resistances. Demand for new types of goods and services is latent, but GM and GE and GF may not be able to produce these things. This is at least a possibility. We represent the first society in the history of the world where life is being shortened by people eating too much. In all other societies, including many today, people still die prematurely because they eat too little. This looks like a qualitative change.

For example, Continental Can recently put into its labor contract three months vacation for people with fifteen years of service. This means that expanded recreational opportunities will be part of the new demand. To supply good recreational opportunities will require the development of the public domain, in part. That does not mean that private enterprise cannot eventually enter this field, but a series of preconditions are required before this range of services can be readily expanded.

The success of the automobile industries, the real success in the United States, was largely the result of a political fluke. It was due to the fact that the farmers, who controlled the state legislatures, wanted to reduce their isolation. They found a way of making the people who wanted the roads pay for them, as the roads were constructed, through a gasoline tax. It was this combination of private and public parallelism that made possible the fantastic success of Detroit. Detroit could have been as able as possible, but without the parallel public development, we would have had no significant automobile industry in the United States. This country would not have run a private highway system.

The consumer wants and can pay for good interurban transit or suburban transit. But the combination of our political structure and interests makes it very hard to mobilize the capital investment and the enterprise—and therefore eventually the employment—which would make a significant forward advance in this area.

A simple answer to this problem would be to capitalize government. We are actually doing that now. The Port of New York Authority is an example. We finally obtained thruways by capitalizing government. The City of New York built the new ballpark for the Mets on a long-term lease.

Let us introduce some considerations that are lacking in many discussions of national income. The fastest growing industry with regard to employment in the United States, other than government itself, is medical services. Medical services grew in the United States without much assistance from private enterprise. They grew through a series of peculiar interlocking enterprise forms known as nonprofit—Blue Cross, Blue Shield, and voluntary hospitals. It was the existence of an embryonic enterprise structure together with the desire of people to spend their money this way that permitted this expansion. We should not assume that money flows are not related to the nature of the investment and the nature of the output. This is as great an error as that made by the classical economists who left money out of consideration.

In the absence of enterprise structures to stimulate new demand, incomes will go down to a level where people will spend all they have, but there would not be the investment that is needed each year to provide jobs for all who are available for work.

FACETS OF THE LABOR MARKET

Today, about 6 percent of the labor force is unemployed. In 1937, which was a submerged business cycle peak, our unemployment rate was over 10 percent. One facet of the unemployment problem is that Negroes have a higher unemployment rate than white workers, and that young people have higher unemployment rates than the average. This has been true for as long as we have records.

There is no doubt that the unemployment rate in the United States has been unsatisfactory. But this unsatisfactory unemployment rate may be the result of factors other than automation and technology. Perhaps it is the result of erroneous economic policy.

Today, researchers in this field are trying to establish a sounder basis of dealing with these problems, by getting a clearer idea of their magnitude, their character and their causes. The tendency to exaggerate by appealing to electronic computers and to make extrapolations about employment based on the most advanced technology does not provide a sound basis for dealing with very real human problems.

During periods of rapid technological change, unemployment has been very low. A period of rapid technological change means high levels of investment, which mean a high level of national income and a low level of unemployment. These relations are not simple; however, the slowing down of technological change with reduction in outlets for investment has been one cause of unemployment in recent years. Coupled with this are problems that have arisen because of the restraints imposed by our international trade relationships.

Some further distinctions must be made. There can be a decline in productivity and a serious unemployment problem arising by virtue of changes in the supply of the labor force. There must then be a distinction made between changes in productivity and the structure of the labor market. A significant part of the present difficulty has to do with the unique shape of the labor supply today, quantitatively as well as qualitatively. We have noted that estimates of growth in terms of productivity do not seem to account for the present level of unemployment. It is probable therefore that other factors are responsible for new orders of trouble.

One impressive fact is that there has been no net gain in employment in the whole manufacturing sector in the last decade. Since, in addition, agriculture is not an expanding area of employment, the question of the preconditions for employment increases in the service sector becomes a crucial question. Our basic model has been expanding employment in manufacturing. This is what we know about. That is what our indices tell us most about. There are, then, important social and economic discontinuities which should be distinguished from the acceleration of technological change.

Recent growth rates considered apart from unemployment would indicate that the last two years have been the most shining example of economic success in the United States in a very long time. The growth rate in GNP has approximated between 6 and 7 percent.

However, our economic system is based predominantly on the assumption that there will be jobs for everybody, that these jobs will provide everybody with income, and that in this way people will be able to get a fair share of the resources available in the United States. Those who say that radical changes are required base their contention on the fact that we can no longer provide jobs for a substantial and growing portion of the number of people in the United States. At the lower end of the skill scale, jobs simply cannot be provided. This is why, it is contended, we need rapid major institutional change.

There are also qualitative problems resulting from the structure of the labor force. The labor force is nonhomogenous. We may be arriving at a Balkanization of the labor force. This again has nothing necessarily to do with technological change per se. The rates of technological change could be declining and there still might be trouble of this nature. It may be connected with technology; it may not.

We know that the under-endowed—those with an IQ of less than 90—are going to find it harder and harder to make a productive contribution to the society. Moreover, there are the enormous difficulties of up-educating the under-educated, of providing better than equal opportunity for the slum dweller, for the Negro. It is not only that a greater percent of Negroes are unemployed. It is that political and psychological issues are involved in changing that percent in the light of changes in technology and the growing demand for skills.

THE ROLE OF GOVERNMENT

In the preceding chapter is an implicit postulate that the private economy is the center of enterprise and employment expansion, while government's role here is to facilitate the progress of the private economy.

We have very real problems of adjustment. We have them now; we have had them in the past, and we will have them in the future. They change, over time, and the role of government changes over time. Adam Smith and Jeremy Bentham were quite clear about the fact that the agenda of government depends on the character of the people, the institution, the circumstances, the place and the time. It may well be that the appropriate role of government today is larger than the appropriate role of government fifty years ago, and that it is different in character as well.

The big change in the character of government is the vast increase in defense expenditures. The amount of increase in nondefense expenditures in all government since 1929 has been 3 or 4 percent of the national income. We might disagree about how best the national income should be divided between government and the private sector, but the arguments go beyond this.

One reason that government's role will have to increase is Russia. During the nineteenth century and the beginning of the twentieth century, there was a lot of discussion of the potentialities of socialism, which was the global term used to cover many aspects of reform. The discussions had their effects, some of which were good effects. In the twentieth century we have not merely discussions of the theory of socialism, but its practice as well. There it stands and it is growing rapidly. Russia, a Communist society, in which there is no private enterprise, no market system, now appears to many to be more productive, to have greater potentiality of growth.

This is a widely held opinion, partly because of propaganda, partly because the Russian economy works better than the laissez-faire and other economists thought it would. As late as the 1920s, many economists, of the Viennese school for example, contended that it would be impossible to operate a socialist society. The existence of Russia has indicated that it is not only possible, but that indeed such a society can expand production—sometimes at terrible costs, of course. This is a fundamental reason for the fact that there is so much more doubt today about the efficiency and effectiveness of a market type economy than, say, in 1900. The Soviet Union has made us aware that we simply cannot tolerate an economy which does not operate on a fairly high level of efficiency, and second, that we will not tolerate the kind of personal costs that were involved in social and technological changes in the past.

There is, also, at least as much pressure from the Common Market and Japan as from the Soviet Union. In systems that are remarkably like our own free-enterprise system, we often see fuller utilization of both human as well as capital resources. Competition from the type of free enterprise under way in Western Europe, with its economic and government planning, poses as great a challenge and threatens our security as much as the more radical socialist forms of organization in the Soviet Union.

There are other reasons for the shift toward a position that there is more for government to do. One of them is the great development of economic theory. Keynes was important in indicating the roles of government denied it before. Today the problem of economic stabilization in the United States is recognized by everybody—we have an Employment Act which was supported by both parties, passed almost unanimously. There is no doubt that the increase in economic knowledge, as well as improvement in our economic institutions, has been an important factor in the increase in government's role.

TRENDS AND SOCIAL CHANGE

Social change in this country does not have an entirely different order of magnitude than fifty years ago. The problem is bigger today largely because our standards as to what is right and decent and proper are higher than they were. The problems that are being generated are no greater; it is that we suffer more when we see them.

There is, however, the point of discontinuity. We have seen a radically new development in the last two or three or four years, and this is likely to continue. There does seem to be an immensely more rapid rate of technological change. There is a division between the age when man and machine worked together, and this age in which the machine works by itself.

There is a profound difference in the work situation today, particularly in terms of people working with other people on a day-to-day basis, whether it is due to technology or whether it is part of the total social picture. There is a growing alienation felt by the workers whether due to the bigness of bureaucracy or the complexity of organization. The lines of communication within unions are cumbersome as well.

However, placing today's workers in historical perspective—and these workers are in heavy industry and light industry, they are white and Negro, men and women—does not indicate cumulative alienation, with higher orders of disturbance today than previously. Quite the contrary. In 1890 men had to work seven days a week to keep their families alive. One of the conspicuous improvements in labor market conditions is the considerably wider degree of autonomy that the workers feel by virtue of the unions and their much higher orders of security.

Perhaps we have assumed too quickly that growth means bigness and bigness means much greater impersonality and much greater complexity; the organization-man image leaps to mind. With bigness do we always get greater impersonality, more inflexibility, and a greater sense of anomie? Perhaps subsystems become created within the big systems, and these in fact let the worker who knows how to move within them become a more free and effective operator than he was in a tight, small group that was in fact much more conformist.

One recent breakthrough is our current interest in what is happening to people. How we measure productivity, or whether we have a national agency for masterminding productivity is important; but so is examining all the other changes in the relationship of the individual to the world in which he lives.

One type of social change that relates to technological change can be seen in what has transpired in the life of a farmer over the past few decades. Twenty-five years ago, let us say, a farmer could have counted on the fingers of his two hands the days he had been off that farm during his lifetime. The road collapsed completely every winter. He could not get in or out of his farm. For a long period of time, he lived at that farm, completely isolated. What were the technological changes that changed his life? One of them was the telephone. This led to an incredible difference in the farmers' social environment. Another was R.F.D. The car and the radio also represented a radical change in the social environment of these people.

These changes involved almost half the population; they were massive social changes, and they were the result of technological change. Technological change has been having a vast impact on society ever since the first changes were made in technology. In communications, for example, we have recently made vast strides. Think what television has accomplished. Now when there is a Cuban crisis, we do not learn about it for the first time a couple of weeks later. The President talks about it to everybody in the entire country at the same time.

There are three examples of noneconomic problems of profound significance that are part of the continuing revolution. In addition, they have an enormous present and potential impact on the style of the society, on the way it operates—not only on its economy—but on the values of people, what they commit themselves to, how they use themselves and their resources.

The first example, which is a direct consequence of certain technological developments, is the radical revision in education which is presently under way, changes in physics teaching, changes in biology teaching, changes in social science, changes in English, use of programmed instruction, use of teaching machines, growth in the size of the college population.

At one time the curricula of the best British universities did not include science. This was taught in trade schools and at night. It was a revolutionary development when universities even offered scientific courses. Today, of course, the enormous stress in education is on science and technology as the proper outlets for creative people. Many continue to plead that more effort should go into these areas even to the detriment of the humanities and to an overemphasis on logic. What this will mean in the long run remains to be seen, but it is a direct consequence of certain technological changes.

The second example which has spectacular implications is the growing role of the scientist as a political agent, the growth of the political scientist, the scientist as a power agent and political manipulator in government. Their concern with how we are to spend $50 billion is a direct consequence of technological change.

The third example is the growing dependency on the use of computers for decision making, because of the computer's capacity to simulate complex environments in a way never before possible, and because the computer provides real time data on a scale never available before. This changes the potency, the potentiality of decision making in many areas. It also changes the obligation of the decision maker vis-à-vis the moral and ethical consequences of his acts, which are or will be more explicit than they have been in the past.

These three profound interactions between technology and society must be studied along with economic measures.

Confrontations and Directions

6   by Eli Ginzberg

Professor of Economics and Director, Conservation of Human Resources Project, Columbia University

A seminar is a continuing dialogue. The statements made reflect in the first instance the knowledge and experience of the members, and second what the members learn from each other as an outgrowth of the interchanges which take place. Inevitably, in the first meetings of a new seminar, much of what is said reflects the preconceptions of the members, which in turn reflect the state of their discipline as well as their worldly experience. It is also inevitable that each member initially presents his basic position and concepts sharply to be sure that as the sessions continue they will receive due consideration. As a result, the minutes of the discussions which follow the first five formal presentations contain viewpoints that are rich, diverse, repetitious, and extreme.

It is the aim of this chapter to extract from these wide-ranging colloquies a group of themes which the consensus held must be directly confronted and evaluated if the seminar is to move from the periphery towards the heart of its concern. Several times during the course of the meetings speakers bemoaned the fact that the group was unduly concerned with what had happened, too little concerned with what might happen. This concluding chapter will be definitely future-oriented. Our objective is to develop an agenda for the next several meetings—an agenda that will enable the Seminar to select the questions it wants to probe in seeking to further its understanding of the complex interrelations between technology and social change.

KEY ASPECTS OF TECHNOLOGICAL CHANGE

Effective discussion depends on at least preliminary clarification, preferably the definition of key terms. The Seminar faced considerable difficulty in sorting out and distinguishing sharply an array of related concepts, all of which are involved in considerations of technology and social change. We early recognized that technology cannot be considered effectively without considering science; and in turn, the trends in science depend in considerable measure on the education and training of scientists, and the opportunities that they have to pursue meaningful and constructive careers. We therefore gave considerable attention to the flow of funds available for research and development, and particularly to the significance of the fact that most of these funds are provided by the federal government in the furtherance of its defense and space missions.

In connection with our search for the sources of technological change, some attention was devoted to whether the United States might soon bump against a talent ceiling, because all those who have the intellectual potential to pursue scientific work are already in the field, or in other fields which require individuals with high orders of mental capacity. We noted that the day may be near when new social mechanisms will have to be developed to allocate the limited number of scientifically talented persons among competing national objectives.

Some consideration was paid to the recent vast increases in the number of scientific personnel and the nevertheless relative infrequency with which significant scientific breakthroughs have occurred. Many felt that a primarily quantitative approach might prove seriously misleading. There was even more concern expressed about the fact that such a disproportionate number of scientists and engineers are engaged in military and space research and development activities which are largely self-contained and do not readily spill over to stimulate the civilian economy.

In this connection, the excessive dependence of the aerospace companies on the federal government and the ways in which government contracting results in the wasteful utilization of scarce scientific and engineering talent came into focus. Concern was also expressed about the difficulties that large corporate enterprises grounded in the civilian economy experience in making effective use of their scientific work force. Some believed that the management of scientists and engineers offer no unique difficulties. If management manages effectively, it can manage research and development effectively. But others saw special difficulties growing out of the conflict between the inherently conservative stance of large corporations and the inherently dynamic orientation of scientists and engineers.

Some discussants felt that managers without special training in science and technology are not able to manage effectively men whose work they cannot understand or appreciate. It was pointed out, however, that one way around this difficulty would be to select for positions of general management more and more men with scientific training. Second, there is room for "translators"—individuals who know enough about both worlds to be effective interlocutors.

There was a general consensus that marked changes are under way in the search for and discovery of new knowledge and in its dissemination throughout the economy and society. No one assumed that this is solely a function of the government and other agencies in society spending even larger sums for research and development and for education, but all agreed that major institutional changes are under way in government, in corporate enterprise, in universities, all of which are caught up in a radically different pattern from that which had prevailed even as recently as the 1930s.

Imbedded in these preliminary explorations about technology are the following questions and themes that the Seminar may want to pursue:

1. Is there any way of measuring the rate of change in scientific activity and its technological consequences? Are there limits which may soon affect the future rate of such activity?

2. What can be learned about the current utilization of scientists and engineers?

3. Is there a danger in the heavy concentration of scientific and engineering talent in the defense-space sector of the economy; and, if so, what policies and mechanisms might be employed to effect a partial redistribution of these resources?

4. Is there a "management problem" in corporate enterprise because many managers do not understand the theories and approaches which underlie the work of their technical staffs? If so, in what directions might solutions be sought?

THE RATE OF TECHNOLOGICAL CHANGE

A seminar entitled "Technology and Social Change" was bound to include many who would want to explore the widespread social, economic, and other disturbances and distortions assumed to be resulting from an acceleration of technological change and to seek solutions and remedies for them. But a group composed overwhelmingly of academicians and including a considerable number of economists would also include many who would question the premise that technological change is accelerating. Many academicians, because of the nature of their training, would ask for evidence in support of the thesis of acceleration. And economists would be certain to question such a contention since they have been unable to uncover, despite detailed empirical studies, any solid evidence that the rate of productivity, or of economic growth, has been accelerating in the United States or in other advanced industrial societies. They would not say that the rate of growth in productivity or economic growth might not increase in the future. But acceleration—No! When confronted with such a claim, economists would remember, many from personal experience, the Technocrats of the early 1930s.

The Seminar see-sawed back and forth between extreme views on the rate of technological or economic change. There were few middle positions. Some simply disqualified themselves from participating in the argument that started during the first meeting and continued unabated through the last.

A shredding out of the positions would include first the arguments for a contention of acceleration: the tremendous increases in resources, both money and men, that have been invested in research and development and that have been reflected in the corporate revolution; automation; the harnessing of atomic energy; advances in communications, the leap into space—and many other manifestations of a technology that is making large and rapid gains.

When pushed by the skeptics for evidence in terms of rates of economic growth, the proponents of the acceleration thesis retreated to the position that since the revolution is only recently under way, the future alone can provide the unequivocal evidence. They held that the revolution is too recent to be adequately reflected in historical series.

A second piece of circumstantial evidence advanced in support of the thesis of acceleration is the major breakthrough currently under way as a result of the computer revolution which is clearly affecting how people think, how they act, and the types of goods and services that they demand and use. Without the computer there would have been no jet plane, no exploration of space, and no prospect of the multiple breakthroughs in medicine and education that are already on the horizon.

This did not exhaust their armamentarium. The proponents of the acceleration thesis also called attention to the multiple pieces of evidence of economic tension and social conflict that are a direct consequence of the new technological changes. They pointed first to the evidence that unemployment is creeping upward, slowly but steadily, that it has in fact increased during each of the three last business cycles in the United States. They called attention to the forward march of automation which is resulting in the dismissal of large numbers of semi-skilled workers, many of whom have little or no prospect of ever again finding a satisfactory job, especially those in their late forties or fifties. They noted in addition the economic difficulties again facing the poorly educated Negro who had found a toe-hold in the mass production industries of the North during and after World War II.

Still further evidence of acceleration was adduced: the fact that the scientist-technologist has had power thrust upon him both in the business corporation and in the councils of government for the simple reason that he alone, among educated men, understands the new theories and their potential applications. Others in positions of power and influence have little option—at least in the short run—but to permit this, while they seek to work out new mechanisms whereby they and the larger public can control their own future. The fact that a small esoteric group of intellectuals, steeped in the new mysteries, could take over the key decision-making posts both in industry and government is additional proof that changes in technology have gone far beyond any to which the Western world has been accustomed.

Thus evidence was marshalled. Any single piece might not stand close scrutiny, yet the sum could not be ignored.

But some members of the Seminar insisted on doing just

that. They did not deny, for instance, that the nation was investing considerably more resources than previously in research and development, but they considered this to be largely beside the point. They felt that the basic question was whether the additional resources were providing significant results or whether the yield was small. Economists have long been trained to think in terms of a point of diminishing returns from the application of additional resources. Many discussants felt that the heavy investments which had been made in research and development, particularly in the 1950s, had not paid off as expected, and that we might soon see a change with a leveling off or a decline in the rate of such investment.

Significant economic breakthroughs had occurred earlier in this century when investments in industrial research had been almost negligible: the expansion in the electric power and automobile industries were outstanding examples. Each decade has given evidence of continuity and change: for the most part, the improvements that were introduced into the major sectors of industry were minor and contributed only modestly to increases in productivity. But usually a few sectors experienced very rapid changes, such as we are now seeing in electronics. It is easy to select from the great number of industries the few that are undergoing very rapid change and to assume that the entire economy reflects these few. This is easy, but it is an error.

An inspection of the data on research and development and of the state of industrial technology disclosed that the vaunted progress which supported the theory of acceleration was restricted to a few sectors of the economy, primarily those closely aligned with defense and space. The rest of American industry was relatively backward with regard both to the level of investment in research and development and in the physical state of its technology, much of which is outdated, especially when compared to the new plants in Europe and Japan.

In their repudiation of the acceleration thesis, the economists did not deny that the unemployment level is too high, but they pointed out that little over a decade ago—at the end of the 1940s—it had reached a higher level and had caused little comment. In the interim, however, our expectations and tolerances had changed. A good thing, perhaps, but not a change in technology.

As to the inroads of automation, the skeptics were even more skeptical. In some steel, automobile, meat packing, rubber, and other plants, large new machines had been installed and output substantially enlarged, while employment remained steady or actually declined. This is hardly new or startling. It has long been the way of the economy. In fact, the basis of a continued rise in the standard of living is the ability of the economy to turn out more goods and more services while using less labor and capital. Since, over the last two decades, the economy had been transformed from one in which most workers were engaged in the production of goods to one in which the majority (more than 3 out of 5) are engaged in the production of services, it does not appear likely that automation is an unlimited industrial threat. After all, it would be difficult, at least in the foreseeable future, for automatic machines to cook meals, cut hair, perform appendectomies, educate young and old, and take over the very large number of other tasks that currently can be performed only by talented or skilled or even unskilled persons.

The skeptics were equally unimpressed with the claim that the enthronement of the scientist-technologist in the centers of power is further proof of a technological revolution. The private manager and the public servant must in one way or another become sufficiently familiar with the competing claims for resources, for manpower and capital remain scarce. It is the duty of the manager in private and public life to weigh the relevance of conflicting claims. There is no way that this task can be transferred to the scientist, even if private entrepreneurs and the Congress were so inclined. Moreover, there is no evidence that those with power and responsibility plan to abdicate. Only technical decisions and expert guidance fall in the domain of the scientist specialist. The entrepreneur in search of profits, or a democracy in search of security, cannot walk away from the task of making decisions about how to use scarce resources.

This, then, was the confrontation between two divergent points of view about whether technological change has been accelerated. What questions remain and how can they best be resolved?

1. How can technological change be measured? Is there any way of gaining additional information about the potentialities of the computer revolution? Is there any way of comparing technological change over two time periods—for instance 1900 to 1929, and 1940 to 1963?

2. Can automation be operationally defined and can any estimate be made of its progress to date and its future potentialities? Is the heavy emphasis of the American economy on services a barrier to its rapid expansion?

3. In light of the very large-scale investments in research and development in recent years, how can we explain that the rate of technological change has not been more rapid? Is it likely that the level of expenditures for research and development will level off?

PRODUCTIVITY AND ECONOMIC GROWTH

As we have seen, the economists led the debate against the theory of acceleration in technological change. Much of the work in their discipline in the last few years has been concerned with considerations of productivity and economic growth. Growthmanship has become a political football as well as the academicians' plaything. Each year has seen the able students in economics become ever more cautious about the ability of the American economy to increase significantly its annual rate of growth. Among others, Denison has demonstrated the many and complicated adjustments which are required to make small gains.

The question about whether changes of various sorts should be measured in absolute or relative terms was touched upon from time to time but never thoroughly explored. A further aspect of the problem was mentioned but escaped careful assessment. Some changes may cumulate and others may not. The impact of changes on the economy and the society will differ substantially depending on whether they are transient or permanent, whether they have a restricted or a multiple effect. These and other ramifications of the many facets of change still remain to be explored.

It early became clear that the impact of technology on contemporary society could be assessed only if some clear referents were identified and a way found to measure them over time. For this reason, the seminar repeatedly found the economists attempting to limit the terms of the discussion to productivity and economic growth and shying away from social change, where the categories are not specified and cannot be measured.

Others in the Seminar, of course, did not share the economists' background and preconceptions. The fact that certain overall measures of productivity did not show significant increases did not seem to these non-economists, and even to some of the economists, to minimize the significance of technological changes. The fact that not one new job had been added to manufacturing employment over a decade was significant in its own terms—regardless of what was happening to productivity as a whole.

It was acknowledged that the full potentialities of the new

technology might not be reflected in productivity data because of the continued sluggishness of the economy during the last five years. It is a well known fact that productivity figures tend to increase as the economy approaches full utilization of plant and equipment. And the last five years have been years of substantial underutilization.

There are further limitations to measuring technological change solely with reference to productivity trends. At least a third, and perhaps as much as two-fifths, of our gross national product is currently accounted for by the output of the nonprivate sectors—by government and nonprofit institutions. In the private sector the output of goods is constantly diminishing, while that of services is increasing. The productivity data as they relate to government and the service sectors are imperfect —frequently worthless. Since these are the fastest growing sectors of the economy, overall measures of productivity trends are seriously defective. Moreover, it is always difficult to take full account of the changes in the quality of manufactured goods. A 1963 Ford is quite different from a 1940 Ford. Here is still another limitation of the data.

The Seminar found it difficult to distinguish among a series of related but different economic measures: productivity, economic growth, and economic welfare. Time and again discussants inadvertently shifted from one to the other. Economic growth has usually been accompanied by increases in productivity, but it also involves the total number and quality of resources available for investment. Productivity can increase even though the national income, measured in terms of goods and services produced, does not increase—if the number of people available for work or the number of hours that they work should decline. Or, the gross national product can increase substantially over a decade, with little or no increase in productivity, if the size and quality of the labor force is substantially increasing or if the capital resources devoted to producing the output are substantially enhanced.

Just as we found it difficult to develop estimates of productivity, so we found it difficult to develop good estimates of national income—particularly at a time when so many items that are being produced and consumed fall outside of the competitive market where the price mechanism could measure them.

Moreover, the big jump that many discussants made between technological change and economic growth could not be justified. It is not impossible, surely not in the short run, for rapid technological changes to lead to serious overcapacity in the durable goods sector of the economy, with a consequent depressive action on the economy as a whole that would slow, or even stop for a time, economic growth. In fact, the years after 1957 give considerable evidence of this type of linkage.

But there is also considerable validity to the proposition advanced by several members of the Seminar that the American economy is currently performing unsatisfactorily because of the slow rate of technological change. Since extraordinarily large numbers of young people are becoming available for work, the only prospect of their being employed would follow upon substantial investment in new industries as well as the expansion of existing industries. One point, and only one point, came clear: there is more than one path from technological change to economic growth. One could not subsume all cases under one generalization.

A further source of confusion which plagued the Seminar from time to time grew out of the interchangeable use of the terms "economic growth" and "economic welfare." The touchstone of economic progress is an increase in the amount and quality of the goods and services available *per head* of the population, not total output. There was general agreement that over the long run technological change had contributed greatly to both economic growth and economic welfare. The real income per capita in the United States has been rising over many decades, in fact centuries, and much of the credit for this was ascribed to the advances in technology.

Some in the Seminar believed that the potentialities of modern technology are so great that, if institutional adjustments could be made, technology could now provide a satisfactory standard of living for everybody in the country, and could also make a much greater contribution than hitherto to raising the standards of living of less prosperous nations. This position, simply stated, is that affluence has superseded scarcity. All that we have to do to benefit from the changed circumstances is to unleash the full productive powers of the new technology.

No single position advanced in the Seminar led to more acrimonious debate. The majority of the group believed that this position was vastly overdrawn and most of the economists considered it totally fallacious. They were impressed with the fact that many millions in the United States are living in poverty and that other families high in the income scale continue to show an appetite and desire for more goods and services. A 4, or even 5 percent annual rate of economic growth is not likely to meet all these domestic needs and desires for many years to come.

Another viewpoint maintained that as the American consumer becomes more affluent, as a consequence of increasing urbanization among other factors, he needs goods and services that cannot be readily supplied by the simple expansion of the private, profit-seeking economy. Many consumers want and need better inter-urban transportation, better recreational facilities, more and better access to health and educational services, and many other services that conventionally fall in the public domain. But with governments unable to increase their tax resources substantially, these needs remain unmet, with consequent losses in employment and welfare. Some discussants saw the effective resolution of these institutional issues as more important than the acceleration of technological change.

These then are some of the issues concerning productivity and economic growth and economic progress that require clarification:

1. To what extent do prevailing measures of productivity, economic growth, and economic welfare provide adequate criteria for assessing the impact of technological change on the economy? What other measures might be useful?

2. To what extent should attention be focussed on absolute or relative rates of change? Can useful distinctions be made between changes which are cumulative and those which are not?

3. If the focus of inquiry is on the economic consequences of technological changes, should it be on specific industries, sectors, or on the economy as a whole? What types of questions could best be answered by what approaches?

4. Is it possible, or even desirable, to deal with the economic impact of technological change without simultaneously considering the other forces affecting productivity, growth, and progress? To what extent is it possible to identify and measure these other forces?

SOCIAL CHANGE

The considerable attention that the Seminar devoted to assessing the impact of technological change in terms of economic referents reflected the belief of many members that these referents are the only ones that can be delineated and measured. All else is shadowy and speculative. But the majority did not in fact adhere to this point of view. Indeed, it was denied by the title of the Seminar which explicitly included a concern with "social change," which includes economic change but goes far beyond it.

Much of the methodological spinning reflected this cleavage between participants who wanted to limit the discussion to the economic consequences of technological change in the hope of pinpointing at least a few elements, and those who sought to roam farther afield even at the risk of getting lost and of pinning down nothing. It was this difference in orientation which led one discussant to remark at the last session that difficult as it is to measure economic change growing out of technological advances, this is still easier than to focus on the much broader area of social change which practically defies definition and measurement.

The composition of the Seminar—which, in addition to economists, sociologists, political scientists, and other social scientists, included participants from science, engineering, philosophy, history, architecture, government administration and business—assured that the proponents of a broad construction would win out. Such a heterogenous group, each of whose members by his presence demonstrated an interest and concern with the problem, could not be satisfied with a focus limited solely to the relationship of technology to economics.

In fact, the opening paper set wide boundaries. Dr. DeCarlo raised a host of questions which could be answered only by considering the impact of the new technology on government, defense, science, the computer, the intellectual, artistic, and moral stance of the society, education, careers, and still other aspects of contemporary American life.

To follow a simple escalation approach the Seminar had occasion to note, although it did not probe, the way in which the new science and technology was likely to affect the education, training, work, and leisure of the average citizen. Stress was laid on the fact that our society would soon have no place left for the unskilled worker, and that even the skilled worker would have to undertake repeated training to keep his skills from obsolescing. While there was general agreement that the general thrust of the technological advances was to place a premium on intellectual work and to threaten the employment prospects of those of limited education, a warning was introduced against exaggerating this trend. There is much work, particularly in the service fields, that apparently continues to require many people of modest skills—from serving meals to caring for the large numbers of the mentally ill and the aged.

A related suggestion was that since scientists cannot qualify to act as managers simply on the basis of their knowledge of science, and since managers cannot avoid making decisions that increasingly involve them in judgments about science and technology, one way out of the current dilemma is to broaden the educational base of both groups—all who attend college should acquire some knowledge of science and all engineers and scientists should have some solid grounding in the humanities and the social sciences. It was further pointed out that even if these educational reforms are made, the leaders of the scientific community still have a responsibility to play a much larger role in interpreting the choices that our nation faces and in making this interpretation, to observe the greatest rectitude.

Considerable concern was expressed at many points in the discussion about the substantial changes that are being made in the structure and functioning of the major institution of the private economy—the corporation. In addition to the questions already mentioned as to whether industrial managers are capable of managing enterprises that are increasingly dependent upon the advances of science and technology in its more esoteric manifestations, attention was directed to the loss of initiative which seems to characterize the large corporations which are heavily involved in government contracts, particularly in the aerospace industry. Much evidence was adduced to the effect that the officials in the Department of Defense, NASA, the Atomic Energy Commission, pursued policies that surely restrict the scope for freedom and initiative of government contractors.

Many participants felt that important results would accrue from studying in some detail the impact and import of these new relations between government and business growing out of the significant scale of defense expenditures.

Several discussants asked whether the nation is deriving benefits in the civilian sector of the economy from the very large research and development expenditures that are being directed to defense and space. There was general agreement that the spillover is very modest and that it would be only a slight exaggeration to say that the civilian sector is being "starved" for research funds, and even more importantly, for research personnel, who are overwhelmingly attracted to the more exciting work on the frontiers of defense and space. It was not clear, however, what mechanisms could be developed to insure that adequate personnel and funds would be devoted to urban renewal, housing, recreation, and the conventional areas of the private economy, particularly manufacturing, which had long held the key to economic progress.

The order of difficulty that our society faces in seeking to make the most of the potentialities inherent in science and technology was suggested by the discussion covering the need for a "systems approach" to take full advantage of these potentialities. But it was suggested that our long-standing preference for the nurturing of competitive enterprises and our distaste for cooperative action among large corporations created a real block to the effective development of a "systems approach." We apparently can have more competition or more exploitation of technology, but it would be difficult—and perhaps impossible—to have more of both simultaneously.

The antitrust laws are only one block. In housing—a basic area in which the nation continues to confront many urgent needs—an elaborate structure of small enterprises, strong unions, local ordinances, restricted financing, conspired to place major hurdles in the path of rapid technological development of the industry. Here is another conflict area between the institutional fabric and the potentialities of technology.

Failure to find solutions in these areas carries with it the heavy cost of slower economic growth and progress, but much more serious from the point of view of most members of the Seminar are the challenges that our society faces in any attempt to modify the institutions so that the new technology would not jeopardize but would strengthen our democratic principles and purposes. Many participants felt that if present trends in scientific and technological developments proceed unchecked, before long a relatively small elite will be in possession of most of the decision-making apparatus in business and in government, and the citizenry will no longer be able to participate in making the crucial choices which will shape their future.

This somber view was challenged by others who insisted that all that is transpiring to alter the shape of things is occurring within a political framework, for this is the nature of our society. There is no reason to fear that the democracy, which has proved itself repeatedly to be so resilient, will collapse under the challenge of the new technology.

It might not be easy for the average citizen to express himself with respect to these esoteric matters; it might not be easy for Congress to gain effective control over the decision-making apparatus with respect to science and technology. But there is no reason for despair. The citizen can understand the values posed by alternative scientifically precipitated policy issues, such as a test ban, even if he remains ignorant of the theories and methods by which they can be realized. At least so it seems to those who are impressed but not overwhelmed by big science. But all agreed that the problems are complex and that they have not attracted the attention and the study that they urgently require if sound solutions are to be developed.

Among the social changes that the Seminar identified as worthy of further study are the following:

1. What do the rapid advances in science and technology imply for education and training at every level?

2. To what extent is the dominance of the private corporation and the private economy being eroded by the large-scale participation of government in research, development, and procurement in many important sectors of the economy? What adjustments in mechanisms for improved planning, operations, and control are called for?

3. Can any meaningful generalizations be ventured about the impact of the new technology on the quality of the life of the individual citizen, his family, and the community? What adjustments, if any, are required so that he can benefit more broadly from the potentialities of the advances in science and technology?

4. What adjustments are called for in the political realm to insure that the citizenry will have the information and guidance necessary to exercise intelligent choices about the future shape of its society?

This chapter is a very abbreviated summary of the wide diversity of opinions that were expressed during the first meetings of the Seminar. It was made selective in the hope that this would enable the participants and others to see some of the more important themes that were identified even if they were not fully explored. New knowledge and improved judgments do not come easily or quickly, especially if the angle of inquiry is broad and the subject matter complex. This chapter underscores more the confusion than the clarifications that were achieved, but the recognition of the nature of disagreements and the specification of issues imbedded in them is a sound and tested way of making progress. The Seminar is now in a position to take the next steps ahead. Improved communications is one important consequence of the advances in technology. A seminar on technology and social change should take advantage of this facet of progress and seek to contribute to and benefit from it.

Members of the Seminar, 1962-1963

William O. Baker
Bell Telephone Laboratories, Inc.

Arnold Beichman
Electrical Union World

Daniel Bell
Department of Sociology
Columbia University

Charles R. Bowen
International Business Machines
  Corporation

Frank H. Cassell
Inland Steel Corporation

Peter Caws
Carnegie Corporation of New York

Neil Chamberlain
Department of Economics
Yale University

Edward Chase
Cunningham and Walsh

Harold F. Clark
Teachers College
Columbia University

Thomas E. Cooney
Science and Engineering Division
Ford Foundation

Charles DeCarlo
International Business Machines
  Corporation

Luther H. Evans
International and Legal Collections
Columbia University

Victor R. Fuchs
National Bureau of Economic
  Research

Edwin A. Gee
E. I. duPont de Nemours and
  Company

Eli Ginzberg
Graduate School of Business
Columbia University

Sylvia Gottlieb
Bureau of Labor Statistics
U.S. Department of Labor

Earl D. Johnson
Delta Airlines

David Kaplan
Economics of Distribution
  Foundation

Melvin Kranzberg
Department of History
Case Institute of Technology

MEMBERS OF THE SEMINAR

James Kuhn
Graduate School of Business
Columbia University

Eric Larrabee
Consultant

Robert Lekachman
Barnard College
Columbia University

Donald N. Michael
Institute for Policy Studies

Jacob Mincer
Department of Economics
Columbia University

Lawrence H. O'Neill
School of Engineering
Columbia University

Emanuel R. Piore
International Business Machines
 Corporation

Ithiel de Sola Pool
Massachusetts Institute of
 Technology

A. H. Raskin
The New York Times

Ormsbee Robinson
International Business Machines
 Corporation

Mario Salvadori
School of Engineering
Columbia University

Donald F. Shaughnessy
University Seminars
Columbia University

David Sidorsky
Department of Philosophy
Columbia University

Adolph Sturmthal
Institute of Labor and Industrial
 Relations
University of Illinois

Frank Tannenbaum
University Seminars
Columbia University

Robert Theobald
Consultant

Henry Villard
Department of Economics
The City College

Charles R. Walker
Yale University

Kirby Warren
Graduate School of Business
Columbia University

Aaron W. Warner
Department of Economics
Columbia University

Gerald Wendt
Consultant

Lawrence Williams
School of Industrial Relations
Cornell University

Seymour Wolfbein
Office of Manpower, Automation,
 and Training
U.S. Department of Labor

Christopher Wright
Council for Atomic Age Studies
Columbia University

Bei Fragen zur Produktsicherheit wenden Sie sich bitte an:
If you have any questions regarding product safety,
please contact:

Walter de Gruyter GmbH
Genthiner Straße 13
10785 Berlin
productsafety@degruyterbrill.com